Biology of Fishery Resources

Xinjun Chen • Bilin Liu
Editors

Biology of Fishery Resources

Springer

Science Press
Beijing

Editors
Xinjun Chen
College of Marine Science
Shanghai Ocean University
Lingang New City, Shanghai, China

Bilin Liu
College of marine sciences
Shanghai Ocean University
Shanghai, China

ISBN 978-981-16-6947-7 ISBN 978-981-16-6948-4 (eBook)
https://doi.org/10.1007/978-981-16-6948-4

Jointly published with Science Press
The print edition is not for sale in China (Mainland). Customers from China (Mainland) please order the print book from: Science Press.

The translation was done with the help of artificial intelligence (machine translation by the service DeepL.com). A subsequent human revision was done primarily in terms of content.

This Springer imprint is published by the registered company Springer Nature Singapore Pte Ltd.
The registered company address is: 152 Beach Road, #21-01/04 Gateway East, Singapore 189721, Singapore

Preface

Fisheries resources are an important component of natural resources and source of human food, and they provide employment, economic benefits, and social welfare to people engaged in fishing activities. Fisheries resource science is one of the major branches of fisheries discipline, and the results of fisheries resource science will provide a scientific basis for the reasonable production of fisheries and management of fisheries resources. Fishery resources biology is a natural discipline that studies the ecology of fish resources and other aquatic economic animal groups, which includes population composition, age and growth characteristics, sexual maturity and reproductive habits, and feeding and migration distribution. Therefore, this discipline lays a solid foundation for future work in marine fisheries production and resource management, teaching, and research.

This monograph is divided into seven parts. The first part describes the concepts of fishery resources and fishery resources science, and the basic characteristics of fishery resources, and briefly expounds the basic concepts, the status of the discipline, the basic research content, and the relationship with other disciplines of fishery resources biology. The second part introduces the formation of the population as well as its basic concepts and characteristics, and discusses the population structure and its changing rules. Combined with the latest research results at home and abroad, the method of population identification is introduced in detail. The general law of population growth and its influencing factors are analyzed. The third part describes the life history of fish and the division and characteristics of the development period in detail, the classification of the life history types of fish, the types of fish eggs, and the morphological characteristics and identification of larvae and juveniles; the environmental factors affecting larval survival are also analyzed. In the fourth part, the general principles of annuli formation and general materials for aging are described in detail. The structure, types, circulus characteristics of scales, and the matters needing attention in collecting scales are also introduced. At the same time, other identification materials such as fish otolith and aging methods, including crustacean aging and cephalopod aging methods, are also described in detail. The fifth part describes the process of fish sexual maturation, the method of gonad maturity division, breeding habits and fertility measurement, and the fish reproductive strategy and its relationship with the environment. The sixth part focuses on the relationship between fish and the food chain, and the types and characteristics of fish feeding and research methods; it explains how fishes ensure their food supply, and at the same time, the conception and research method of fatness and fat content are introduced. The seventh part describes the concept of fish schooling as well as the cause of formation and the type of fish schooling, discusses the structure of fish schooling and the law of its change, and introduces the concept, the type, the mechanism, and the biological significance of fish migration. At the same time, combining the latest research results at home and abroad, this part introduces the research methods and cases of fish migration, which provides a basis for understanding and mastering the distribution of fish schooling and migration.

This monograph will provide a reference material for those who want to engage in fisheries or fishery-related work, which is applicable to undergraduate, postgraduate, training education, and fishery-related personnel. Meanwhile, due to the limitations of length and reference materials, as well as the limited level of authors, there are still many inappropriate points in this monograph. Readers are requested to provide corrections and suggestions.

Shanghai, China Xinjun Chen

Contents

Contributors

Xinjun Chen College of Marine Sciences, Shanghai Ocean University, Shanghai, China

Zhou Fang College of Marine Sciences, Shanghai Ocean University, Shanghai, China

Yunkai Li College of Marine Sciences, Shanghai Ocean University, Shanghai, China

Dongming Lin College of Marine Sciences, Shanghai Ocean University, Shanghai, China

Bilin Liu College of Marine Sciences, Shanghai Ocean University, Shanghai, China

Check for updates

Introduction

1

Xinjun Chen

Abstract

Fisheries resources are an important component of natural resources and a source of human food, and they provide employment, economic benefits, and social welfare to people engaged in fishing activities. Fisheries resource science is one of the major branches of fisheries discipline, and the results of fisheries resource science will provide a scientific basis for the reasonable production of fisheries and management of fisheries resources. Fishery resources biology is a natural discipline that studies the ecology of fish resources and other aquatic economic animal groups and is a branch of biology. It is a science that developed gradually in conjunction with human production activities for fisheries production and is the development of ichthyology and aquatic zoology and their practical application in production. Fishery resources biology includes population composition, age and growth characteristics, sexual maturity and reproductive habits, feeding and migration distribution, etc. It also includes an overview of the world's major fishery resources, as well as a grasp and understanding of survey methods of fishery resources and oceanography. Therefore, this discipline lays a solid foundation for future work in marine fisheries production and resource management, teaching, and research. In this chapter, we mainly introduce the concepts of fishery resources and fisheries resource science, and the basic characteristics of fishery resources, and briefly expound the basic concepts, the status of the discipline, the basic research contents, and the relationship with other disciplines of fishery resources biology, and the significance of studying fishery resources biology was also analyzed.

Keywords

Fishery biology science · Fishery resources · Basic concept · Research content

Abbreviation

FAO Food and Agriculture Organization

Fisheries resources are an important component of natural resources and a source of human food, and they provide employment, economic benefits, and social welfare to people engaged in fishing activities. According to the Food and Agriculture Organization (FAO) of the United Nations (FAO 2020), fish products accounted for 17% of the animal protein and 7% of the total protein consumed by the global population in 2017. In addition, fish account for almost 20% of animal

X. Chen (✉)
College of Marine Sciences, Shanghai Ocean University, Shanghai, China
e-mail: xjchen@shou.edu.cn

X. Chen, B. Liu (eds.), *Biology of Fishery Resources*, https://doi.org/10.1007/978-981-16-6948-4_1

protein intake for over 3.3 billion people, with the share reaching 50% or more in countries such as Bangladesh, Cambodia, Gambia, Ghana, Indonesia, Sierra Leone, Sri Lanka, and some small developing island nations (FAO 2020). Fish remain the most traded food product in the world. In 2018, 221 countries reported exports of fishes and fisheries products, with 38% (67 million ton) of total fisheries products and aquaculture production entering international trade. Fish and fisheries products accounted for approximately 11% of the total exports of agricultural products (excluding forest products). Many of the world's people rely on fisheries and aquaculture as a source of income and livelihoods, with 59.51 million people primarily engaged in capture fisheries and aquaculture in 2018 (FAO 2020).

In China, the position of fisheries in the national economy has been increasing. According to national statistics, in 2019, the total amount of aquatic products in the country reached 64.8 million tons, of which domestic marine fishing output totaled ten million tons and offshore fishing output totaled 2.17 million tons. The total number of fishing vessels was 731,200, with a total tonnage of 10.424 million tons. The total output value of fisheries in society as a whole is 2640.650 billion yuan (current year price). The fisheries population is 18,282,000, and 12,917,000 people are employed in the fisheries industry. The total output and value of import and export of aquatic products together are 10.5332 million tons and 39.359 billion US dollars. The national per capita consumption of fisheries products is 46.45 kg. Thus, fisheries resources play an important role in food security, employment of fishermen, economic development, and foreign trade.

1.1 Basic Concepts and Characteristics of Fisheries Resources

1.1.1 Concept of Fisheries Resources

Fisheries resources, the material basis for the development of aquaculture, are also important sources of human food and high-quality animal protein. Fisheries resources are a general term for the species and quantity of economic animals and plants (fish, shellfish, crustaceans, sea animals, and algae) in natural waters that have exploitation value (Chen and Liu 2017).

Fisheries resources are diverse, with the main categories being fish, crustaceans, mollusks, algae, and mammals, and the numbers within each group vary considerably. Fish are the most abundant group of fisheries resources; crustaceans mainly refer to shrimp and crabs; mollusks mainly include shellfish and cephalopods; cephalopods include Ommastrephidae, Loliginidae, Sepiidae, and Octopodidae; and algae include kelp, nori, and other types.

Marine fisheries are the most abundant of all fisheries resources. Depending on the water level, they can be divided into (1) demersal species (mainly cod and hake), which mainly inhabit the bottom layer and are usually trawled, accounting for more than 40% of global marine fisheries production; (2) rocky reef species (such as grouper), which inhabit the rocky reef area and are mainly caught by jigging; (3) coastal pelagic species, which are found in the pelagic zone of the continental shelf and include mainly herring, anchovy, scads, and mackerel; and (4) oceanic pelagic fish (such as tuna), which inhabit the surface layer of the continental slopc and the translucent layer of the oceanic zone.

1.1.2 Concept of Fisheries Resource Science

Fisheries resource science is one of the major branches of aquaculture. Fisheries resource science involves the study of natural life history processes such as population structure, reproduction, feeding, growth, and migration of fish and other species; the study of the population size change patterns of fish and other species, as well as the estimation of resources and catchable amount, the population size change pattern, and its uncertainty under different management strategies; the study of resource exploitation and utilization of fish and other species and socioeconomic development and the law of optimal resource allocation; and the study of fisheries resource management and conservation measures and other topics to provide a basis for the reasonable production of fisheries and scientific management of fisheries resources in science (Chen 2014; Chen and Liu 2017).

1.1.3 Basic Characteristics of Fisheries Resources

Fisheries resources are a type of natural resource with distinct natural characteristics. Unlike inexhaustible natural resources, such as tidal energy and wind energy, and completely different from exhaustible but nonrenewable natural resources, such as minerals, these resources are a renewable biological resource, and most species are mobile and seasonally migratory across regions and large areas, so fisheries resources have unique properties. In-depth analyses and studies of the natural characteristics of fisheries resources are of great importance to the sustainable development and utilization of fisheries resources, scientific management, etc. Fisheries resources have the following natural characteristics (Chen 2014; Chen and Liu 2017):

1. Renewability. Fisheries resources are a renewable resource with the ability to reproduce on their own. Through the reproduction, development, and growth of a population, resources can be constantly renewed, the population is constantly replenished, and the population is balanced at a certain point through a certain self-regulation ability (due to the restrictions of prey, space, and enemy constraints). If suitable environmental conditions are available and human exploitation is reasonable, then fisheries resources can reproduce for generations and continue to provide high-quality animal protein for humans. However, if the environmental conditions for growth are destroyed by natural or man-made damage (e.g., destruction of fish spawning grounds due to land reclamation or habitat changes due to El Niño) or by indiscriminate fishing by humans, then the self-renewal capacity of fisheries resources is reduced, and the ecological balance is disrupted, which will lead to the decline or even depletion of fisheries resources in the long term.
2. Migratory or mobile. With the exception of a few sedentary and sedentary aquatic organisms, the vast majority of fisheries resources have a habit of migrating and moving through the water, which is one of their most distinctive features from those of other natural resources. In economic terms, this mobility is also the fundamental reason why natural property rights are difficult to fix or define and is one of the main causes of overfishing. For example, in general, crustaceans have a smaller range of movement, while fish and mammals have a larger range of movement, especially pelagic fish, including anadromous spawners such as salmon, and tuna. Many species migrate to inshore marine areas when spawning and swim out to sea after spawning. Many species live in different sea areas at different stages of development. As a result, a number of fisheries resource stocks inhabit waters under the jurisdiction of multiple countries or areas throughout their life cycle.
3. Sharedness. With the exception of territorial seas and the 200-nautical-mile exclusive economic zone, a large part of the ocean is not divided into national boundaries. Even in the case of a country's territorial sea or a river that crosses a region, there are generally no obvious boundaries, such as provincial or state boundaries. For migratory species, there are even fewer national and regional boundaries. Because of the seasonal migratory and mobile nature of fisheries resources, a particular fishery resource, or even the same stock, is often the subject of joint exploitation by several countries or regions in a given body of water. It is difficult to confine fisheries resource management to a particular sea area, and similarly, a particular fisherman cannot prevent others from coming to fish. Thus, in economic analyses, fisheries resources are usually

characterized by nonexclusivity of use or consumption. People are free to enter the fishery, and property rights to the fishery resource are often determined at the time of catch by the fisher. This is a typical shared resource.

Shared resources are nonexclusive, and nonexclusivity is a lessening of property rights that will lead to inefficient management and use of resources. In this case, prices cannot act as a coordinator among users for the allocation and use of a resource or provide incentives for its production or conservation and for increased incomes. In resource economics, it is argued that the ultimate allocation of a fishery resource results in overexploitation of the resource and significant underinvestment in resource management, conservation, and enhancement of its productive capacity.

4. Perishability of fish catches. Fisheries resources, such as fish, are an important source of high-quality animal protein for humans. However, if a catch is perishable, then it will completely lose its utility value and use value; even if it is not perishable, if the freshness decreases, then the utilization effect of the fish products will be reduced. Raw tuna is usually bled and gutted immediately after being caught on board a vessel, and then, it is frozen and preserved at −60 °C to ensure its quality; in contrast, if tuna undergoes a general freezing treatment, then its quality will be drastically reduced, and the price of catch will drastically decrease. Therefore, in the era without preservation measures, the scope of fisheries utilization and distribution was greatly restricted, fisheries production was limited to coastal waters, and fisheries product consumption was also limited to coastal areas. With the development of freezing technology and ultralow temperature technology (−60 °C), the expansion of operational fishing grounds to pelagic areas and the storage of large quantities of processed raw materials have been promoted, creating good conditions for the development of fisheries production and the extension of the industrial chain, as well as the large-scale development and utilization of offshore and pelagic fisheries resources.
5. Volatility. Fish and other aquatic animals are cold-blooded species. Therefore, the growth and death of fish and other fisheries resources are highly susceptible to natural factors such as climatic conditions and the hydrological environment, in addition to the effects of man-made fishing activities where there are more unpredictable factors and the amount of resources is more volatile from year to year. Abnormal changes in water temperature, currents, and other factors can greatly harm the growth of fish and other species and have a great impact on the amount of fisheries resources, such as the dramatic drop in anchovy production in Peru caused by El Niño. The volatility of fisheries resources has led to great uncertainty and risk in fishing production activities.
6. Holistic nature. Fish and other species are important components of marine ecosystems, and most are in the middle and upper levels of the marine food chain. Therefore, there is a close relationship between fisheries resources and the various natural environmental conditions under which they survive. For example, predator and prey conditions and human production activities are interrelated and mutually constraining, and changes in one resource element or environmental condition will cause corresponding changes in other related resource elements. For example, there are two fisheries A and B, and competing species S_1 and S_2 coexist and live together when there is no exploitation; however, if an increase in the intensity of fishing species S_1 (fishery A) leads to an increase in species S_2 (fishery B), then fishery A produces a positive external effect on fishery B. This type of effect is called an externality under competitive coexistence conditions in resource economics (externality under competitive coexistence).

From the above analysis, it is clear that the quantity of fisheries resources and their distribution not only are influenced by their own biological characteristics but also change according to changes in habitat conditions and the status of human exploitation. In addition, with the increasing progress of human society, science, technology, and means of production, the types of fisheries resources exploited, the exploited sea areas, and the exploited water layers are also expanding.

1.2 Concept of Fishery Resources Biology and Its Study

1.2.1 Basic Concepts of Fishery Resources Biology

Fishery resources biology is a natural discipline that studies the ecology of fish resources and other aquatic economic animal groups and is a branch of biology. It is a science that developed gradually in conjunction with human production activities for fisheries production and is the development of ichthyology and aquatic zoology and their practical application in production. Of the world's fisheries resources, fish are the main object of human exploitation and utilization, and their production is predominant; thus, we tend to consider fish as the main objects of studies on the biology of fisheries resources.

With the development of the discipline, the connotation and extension of fishery resources biology are constantly enriched and differentiated. Therefore, we believe that there are two concepts of fishery resources biology in a broad and narrow context. In the broad context, fishery resources biology refers to "the study of population structure, age, and growth; migration and distribution; feeding and reproduction; distribution of fishing grounds and environmental relationships; population changes; resource assessment; and management of fish and other fishery organisms." In the narrow context, fishery resources biology is "a discipline that studies the population composition of aquatic organisms such as fish, and the life history processes and characteristics of populations at various stages of the life cycle of fisheries organisms, including age composition, growth characteristics, sexual maturation, reproductive habits, feeding, and migratory distribution, centered on fish populations." In this book, the biology of fisheries resources is studied based on the narrow context.

1.2.2 Disciplinary Status of Fishery Resources Biology

Fishery resources biology is a basic discipline in marine fisheries science and technology and other majors, is a comprehensive basic applied science that focuses on the biology of fish and other aquatic animal groups, and is a part of applied ecology as well as biology. Because of the extremely broad scope covered by this subject, it is both fundamental and applied and has the nature of an integrated science.

The focus of the fishery resources biology discipline includes the basic theories and skills necessary for scientific and technical personnel engaged in marine fisheries production, management, and research. Through this discipline, one can master the basic methods of studying fish stocks, growth, feeding, and reproduction and other aspects of fish biology, and this discipline lays a solid foundation for future work in marine fisheries production and resource management, teaching, and research and provides scientific methods and means of fisheries production and fisheries resource management and their sustainable use.

1.2.3 Basic Research on the Biology of Fisheries Resources

To sustainably use fisheries resources, it is necessary to be familiar with the reserves of fisheries resources in the water and their distribution and their biological characteristics, such as growth, reproduction, mortality, and migration distribution, which are extremely important research subjects in the marine fisheries discipline. Based on years of practice in fisheries production and abundant information from experiments in fisheries science, those engaged in fisheries science have elevated information about the life, habits, distribution, and migration of fisheries resources into scientific theories and identified systematic laws, which have become an extremely important part of fisheries science.

The main types of information obtained from studying the biology of fisheries resources are as follows:

1. A good understanding of the basic theories and methods related to the biology of fisheries resources, such as stock identification, structure, age, and growth; food habits; abundance; reproductive habits and fertility; fish community structure and its biodiversity; early life history traits of fish; and the characteristics of each stage, to provide essential information for assessing fisheries resources, population changes, and fishery forecasting and understanding fish life history.
2. Knowledge of methods and basic concepts of clusters and the migratory distributions of fishes, for example, general rules and principles of fish clustering, types of fish migrations, and research methods.
3. An overview of the world's major fisheries resources, including tuna, cephalopod, and pelagic fish; however, they are not described in this book.
4. Knowledge and understanding of survey methods for fisheries resources and individual fisheries, mainly including marine environmental surveys, marine biological surveys, and fisheries resource surveys, which are not described in this book.

1.2.4 Fishery Resources Biology in Relation to Other Disciplines

As a specialized discipline formed by the intersection of fisheries science, ecology, biological science, and marine science, fishery resources biology has a very close relationship with many other related disciplines, which are briefly described as follows:

1. Ichthyology. Ichthyology is known as a branch of zoology that studies the morphology, classification, physiology, ecology, and genetic evolution of fish. Since fish are the main object of study in fisheries, they are the basis of the fishery resources biology.
2. Marine biology. Marine biology is a biological science that studies marine plankton and benthic organisms. Since plankton and benthic organisms are closely related to the objects of study in fishery resources biology and provide sufficient prey for the growth of fish, they are the basic subjects of the fishery resources biology discipline.
3. Ecology. This discipline is a science that focuses on the study of the interrelationship between organisms and the environment. Since fishery resources biology is a branch of applied ecology, the basic theories and methods of ecology have become the basic content and core of this discipline and guide the direction of it.
4. Fish behavior. Fish ethology is the study of the interrelationship between fish activity status and environmental conditions, particularly the relationship between water temperature, salinity, currents, and other conditions and the distribution of fish migrations, and it provides the basis for the development and study of fishery resources biology.
5. Oceanography. Oceanography is the study of the hydrographic, chemical, and other inorganic and organic environmental factors of the oceans and the laws of their variability and interaction; therefore, the marine water environment is the vehicle for the study of fishery resources biology, and together with ichthyology, it is the basic subject of this discipline.
6. Fisheries stock assessment. This discipline is a separate science from the fisheries biology component of fisheries stock dynamics; is centered on the study of mortality, replenishment, population dynamics, and resource management of fisheries organisms; and is a development, service, and successor to the major areas of fishery resources biology.
7. Environmental biology. This discipline is a science that has gradually developed and emerged in recent decades as the quality of

the environment has declined and endangered biological germplasm resources and fish. From the perspective of biology and ecology, this discipline focuses on major issues such as conservation biology, biodiversity, and large marine ecosystems and explores the relationship between environmental changes and changes in marine biological resources, thus providing a scientific basis for the maintenance of biodiversity and the sustainable use of biological resources.

In addition, there are disciplines such as physiology, biochemical genetics, stock enhancement resources, biostatistics, and molecular phylogeography, all of which also provide the means and methods that together contribute to moving the development of fishery resources biology forward.

1.3 Significance of Research on Fishery Resources Biology

There are a wide variety of fisheries resources, mainly fish, crustaceans (shrimp and crab), mollusks, algae, and mammals (whales and dolphins), of which there are more than 20,000 species of fish; however, only 130 species of these fisheries resources are the main fished species, with fish accounting for the majority. According to the FAO study, the top 15 species in terms of production are anchoveta (*Engraulis ringens*), Alaska pollock (*Gadus chalcogrammus*), skipjack tuna (*Katsuwonus pelamis*), Atlantic herring (*Clupea harengus*), blue whiting (*Micromesistius poutassou*), European pilchard (*Sardina pilchardus*), Pacific chub mackerel (*Scomber japonicus*), yellowfin tuna (*Thunnus albacares*), scads nei (*Decapterus* spp.), Atlantic cod (*Gadus morhua*), largehead hairtail (*Trichiurus lepturus*), Atlantic mackerel (*Scomber scombrus*), Jack and horse mackerels nei (*Trachurus* spp.), Sardinellas nei (*Sardinella* spp.), and jumbo flying squid (*Dosidicus gigas*), whose cumulative production exceeds 30 million tons, accounting for more than 35% of the global fishing production. In 2000–2018, the total world marine capture production was stable between 78 and 84.4 million tons (FAO 2020).

In addition to being used for human consumption, fisheries resources are of high economic value as feed for marketable animals and as raw materials for industry and medicine. Fisheries resources play an important role in food security, employment of fishermen, economic development, and foreign trade, so ensuring the sustainable use of fisheries resources is an extremely important research topic. To ensure the sustainable use of fisheries resources, it is necessary that the biological characteristics of fisheries resources and their organic linkages with habitats be identified to determine the biological laws that control and modify these biological resources so that services and a basis for the rational exploitation can be provided, fisheries resources can be used by humans, and a marine fishing environment can be created. With the continuous development of science and technology and the further increase in human demand for fish protein, humans are also changing from only fishing to coordinating the direction development and fishing, and therefore, studying the biology of fisheries resources has become increasingly important.

The biology of fisheries resources refers to the basic biological mechanisms of population changes in fish and other economic aquatic organisms, which mainly include population, reproduction, development, feeding, and growth, and population structure, among which the study of population and its structure is the most important, and the scientific delineation of a population is the prerequisite for qualitatively and quantitatively studying resources. Research on the biological basis of fisheries can provide a theoretical basis for the assessment of fisheries resources, fishing forecasts, resource management and enhancement, etc.

References

Chen XJ (2014) Fisheries resources and fishery oceanography. Ocean Press (in Chinese)

Chen XJ, Liu BL (2017) Biology of fishery resources. Science Press (in Chinese)

FAO (2020) The state of world fisheries and aquaculture 2020, Rome

Fish Stocks and Discriminant Methods

Xinjun Chen, Bilin Liu, and Zhou Fang

Abstract

Population is the basic unit of species existence, heredity, and evolution, the basic structural unit of biological community and ecosystem, and the specific unit of fishery resources development and management. Moreover, population identification is the basis of studying the population dynamics and living habits of fish. Only by understanding the population structure of fish can it provide a basis for the sustainable use and scientific management of fishery resources. In the assessment and management of fishery resources, it is usually based on the understanding and mastery of the biological characteristics of fishery population objects, taking certain assumptions as the premise, through establishing mathematical models, to describe and estimate the composition and structure of the population, the amount of resources, and their changes, to evaluate the impact of fishing intensity and fishing specifications on the population, and to grasp the changing characteristics and laws of the amount of resources of the population. Thus, simulation and prediction of the past and future status of resource groups can provide scientific basis for the establishment and implementation of management measures of fishery resources. Therefore, the population is the basic unit of fishery resources research. In this chapter, the formation of the population and the basic concepts and characteristics of the population are described in detail, and the population structure and its changing rules are discussed. Combined with the latest research results at home and abroad, the method of population identification was introduced in detail. The general law of population growth and its influencing factors were analyzed. At the same time, taking large yellow croaker (*Pseudosciaena crocea*) and oceanic squid (Ommastrephidae) in the high seas as examples, the division of different geographical populations and spawning populations and their relationship with environmental factors were analyzed in detail, which provided the basis for understanding and mastering the basic methods of population identification. The focus of this chapter is to master the basic concepts of stocks and related species and to master the methods and techniques of stock discrimination.

Keywords

Fish stock · Discriminant method of stock · Population structure · Large yellow croaker

X. Chen (✉) · B. Liu · Z. Fang
College of Marine Sciences, Shanghai Ocean University, Shanghai, China
e-mail: xjchen@shou.edu.cn; Bl-liu@shou.edu.cn; Zfang@shou.edu.cn

X. Chen, B. Liu (eds.), *Biology of Fishery Resources*, https://doi.org/10.1007/978-981-16-6948-4_2

Abbreviations

AAS	atomic absorption spectrometry
AFLP	amplified fragment length polymorphic
B, *I*, *D*, and *U*	the number of populations being born, moving in, dying, and moving out at time t to t+1, respectively
BS	proton back scattering
C.D.	Coefficient of difference
CVA	Canonical variance analysis
d	the probability that an individual will die in a given time
d_i	denotes the off-mean difference of i population characteristics
dN/dt	the change in population size over a very short time interval
DNA	deoxyribonucleic acid
E	the fishing intensity
EPMA	electron probe microanalysis
EST	expressed sequence tag
h	rate of fishing
HK	geographic population of Haikou, Hainan
HLJ	Heilongjiang population
ICP-MS	inductively coupled plasma-mass spectrometry
ICP-OES	inductively coupled plasma with optical spectrum analysis
ISSR	intersimple sequence repeat
JJ	Jiujiang population
k	mean nucleotide difference number
K_1	the load capacity
$L_{50\%}$	body length at first sexual maturity
LA-ICPMS	laser ablation inductively coupled plasma mass spectrometry
Li	each body length group
M_1 and M_2	denote the mean of the two population characteristic measures
m_1 and m_2	the mean errors of the two population characteristic measures
M_{diff}	significance of difference in means
MSY	maximum sustainable yield
N	the population size
N_0	the population size at moment 0
n_1 and n_2	the sample sizes of the two population characteristics
nanoS IMS	nanosecondary ion mass spectrometry
NC	Nanchang population
N_t and N_{t+1}	the population size at time t and t+1, respectively
PC	principal components
Pi	the percentage of mature individuals in the group sample
PIXE	proton-induced X-ray emission
PT	geographic population of Pingtan, Fujian
Q	the fishing coefficient
QTJ	Qiantang river population
r	the population endogenous growth rate
R_0	the net growth rate per generation
RAPD	random amplified polymorphic DNA
RFLP	restriction fragment length polymorphism
RNA	Ribonucleic Acid
RU	Russian population
S_1 and S_2	the standard deviations of the two population characteristic measures
S_{ij}	denotes the sum of the covariances of i and j population characteristics
SNP	single nucleotide polymorphism
SSCP	single-strand conformation polymorphism
SSRs	simple sequence repeats
S-XRF	synchrotron X-ray fluorescence spectroscopy
TW	geographic population of Taiwan
UPMGA	cluster analysis
XM	geographic population of Xiamen, Fujian
YT	Yingtang population;
YY	Yueyang population;
ZJ	geographic population of Zhanjiang, Guangdong
$\lambda_1, \lambda_2..., \lambda_k$	the judgment coefficients can be solved
π	nucleotide diversity index

2.1 Population

2.1.1 Basic Concept of Population and Its Formation

It is well known that animals are not evenly distributed in nature but live in scattered territories with significant territorial and spatial variation. Such clusters of individuals of the same species within a certain environmental space then gradually form populations. Therefore, this book defines a population as "an assembled group of homogeneous organisms occupying a specific space at a specific time." Thus, a population is a group of individuals in a group or several groups within the distribution area of a species that have similar morphological characteristics, the same physiological and ecological characteristics, and common reproductive habits, i.e., the same genetic attributes – the same gene pool. For example, Pacific herring that live year-round in the Yellow Sea may be referred to as the Yellow Sea stock.

Populations have certain characteristics due to varying geographic distributions, environmental conditions, and life histories among species. In general, the following are three main characteristics of populations:

1. Genetic characteristics. Populations have a certain heredity, i.e., a certain genetic composition, and belong to the same gene pool. Since all populations have their own genetic properties, these properties are the basic unit of population genetics research. On the one hand, individuals can exchange genetic factors among themselves to promote the prosperity of populations, while on the other hand, populations maintain differences in morphological, physiological, and ecological characteristics among themselves.
2. Spatial characteristics. All populations have a certain distribution range within which there are suitable conditions for the survival of the population. The center of their distribution is usually the most suitable, the peripheral areas fluctuate considerably, and the boundaries are often blurred and sometimes crossed.
3. Quantitative characteristics. Population size varies with time and environment, and it has its own inherent pattern of quantitative change, usually with a basic range. Population density and size often vary, even greatly. Each population usually has corresponding life history characteristics, such as birth rate, replacement rate, growth rate, and mortality rate.

2.1.2 Causes of Population Differentiation

One of the basic biological evolutionary processes is divergent evolution. That is, as a result of spaced differentiation of populations, subspecies may be formed, and by successive differentiation of subspecies, new species may be formed. Thus, species formation is the development of a uniform propagule into a new, spaced out propagule. Population differentiation results from the action of isolation mechanisms, which are generally manifested in five ways: geographical, ecological, reproductive, seasonal, and physiological.

1. Geographic isolation. The separation of multiple groups, whether within a contiguous geographic area of existence or with gaps in distribution, exists in different areas, even if their spatial distribution does not overlap and they cannot interbreed with each other.
2. Ecological isolation. Groups survive within the same territory but are each subject to different conditions of existence, and over time, each accumulates different genetic traits to adapt to different biomes.
3. Reproductive isolation. Differences in sexual physiology occur between groups that limit or inhibit reproductive exchange between them.
4. Seasonal segregation. Periods of mating or growth occur in different seasons.
5. Physiological segregation of sexes. A weakness or lack of mutual attraction between the sexes of different species or physical incompatibility of genitalia occurs.

2.2 Population Structure and Patterns of Change

2.2.1 Basic Meaning of Population Structure

Stock structure is the state of heterogeneity in the morphology of fish and their quantitative characteristics within a generation (or a stock), specifically the proportion of the number and biomass of age and length groups within a stock, the proportion of sexually mature fish in a stock, the proportion of older fish to the remainder of the same stock, and the proportion of males and females in a stock as a whole or in each age group or length group.

In the case of fishery resources biology, the main characteristics of stock structure include four aspects: age structure, individual composition (length and weight), sex structure, and sexual maturity composition.

Different fish species and populations have different community structures. Usually, population structure is remarkably stable. However, since populations live in a constantly changing environment, population structure, like other attributes of a population, is constantly changing within a certain range to adapt to changes in the external habitat.

2.2.2 Age Structure and Its Evolution

(1) Concept of Age Structure and Its Connotations

A population includes individuals of all different ages, thus constituting the age structure of the population. Age structure refers to the maximum age, the mean age, and the percentage composition of individuals in each age class of a fish population and is one of the important characteristics of a population.

The age structure of fish populations is related to the length of their life span. Long-lived fish have a large number of age groups, a complex structure, and a multiage structure type. Short-lived fish populations have fewer age groups, a relatively simple structure, and a simple age structure type. Thus, the lengths of the life spans of fish vary, as do their age structure. Different populations of the same fish species have different age structures. This characteristic is an adaptive property of a population that ensures its survival in its specific living environment.

In comparison to populations with a simple age structure, those with a multiage structure have a relatively wide range of prey options and a more stable prey base and more strongly defend adults against enemies, so aggressive animals are less harmful to sexually mature individuals in the population; in addition, the population sexually matures later and has a slower reproductive rhythm. Fish with a simpler age structure have a relatively unstable prey base, are more strongly influenced by aggressive animals, have higher natural mortality rates, have more pronounced population size variability, reach sexual maturity earlier, and have more variable reproductive rhythms.

1. Life Spans of Fish and Their General Characteristics

The life spans of fish vary greatly. Some gobies live only a few months, while some sturgeons can live for hundreds of years. Traditional economic fish tend to live from the age of 2 to a few dozen years.

The life spans of fish vary at different latitudes. In the Northern Hemisphere, for example, fish distributed at midlatitudes tend to be most abundant at ages 5–15 years old and at lengths of 30–50 cm; those distributed in equatorial waters have a lower average life span. In addition, the average age and age range of fish in southern waters are less variable, and sexual maturity is earlier, which is related to the different levels of influence of aggressive animals and to adaptations that ensure a high reproductive capacity of a population.

The life spans of fish with different feeding characteristics vary. In general, the longest-lived and largest fish are generally large, aggressive fish that feed vigorously for a short period of time; benthic-feeding fish, as well as some herbivorous and carnivorous fish, are essentially medium (1 m or slightly larger) and approximately 30 years old; plankton-feeding and small benthic-feeding fish are generally mostly small fish with short life cycles, and almost none are aggressive. There are no fish that are aggressive.

In comparison to migratory populations, sedentary coastal populations are generally characterized by shorter life cycles and smaller individuals. This difference is mainly related to differences in food security and in most cases is not related to the impact of aggressive animals.

Different populations of the same fish species may also differ in age range and maximum body length, reflecting the adaptation of a population to its living environment. The population structures of longer- and medium-lived fish vary considerably, as do those of short-lived fish, but to a lesser extent.

2. Concept and Meaning of the Age Composition of Population

The age composition of a population is the proportion of the number of generations in a population. Changes in the number of generations are the result of the interaction of three processes: population replenishment, growth, and mortality reduction. Changes in age structure, both for the population as a whole and for its sexually mature component, depend on the proportions of these three interacting processes. The number of generations varies and has a direct effect on the age composition of a population. In some fish, the number of strong generations can be tens or even hundreds of times higher than the number of weak generations. The addition of a strong generation inevitably causes a decrease in the proportion of older fish; conversely, with a small number of newly added generations, the proportion of older fish in the population increases relatively.

In addition to changes in generation occurrence that have a significant effect on the age structure of a population, changes in food security and the growth of fish within a population also have a significant effect on the age composition of the population. The length of most fish at first sexual maturity is approximately half the maximum length that the fish can reach. Therefore, if a population has good nutritional conditions, improved food security, and a long prey season, which leads to accelerated fish growth, then it will reach the length range of sexual maturity at a lower age, and sexual maturity will occur earlier, sometimes even causing changes in sex ratios and shorter life spans. For example, in the age composition of northern plaice, in years of intensive fishing, the nutritional conditions of the stock improve, the prey base is strengthened, and the growth rhythm of the fish accelerates, resulting in a lower aging of the reproductive population.

Long-lived populations with multiage structures have complex structures and relatively flat interannual variations in population size. Typically, such stocks are not only composed of a larger number of age groups for their repeat reproductive population but also have a multiage structure for the replenishment population. This scenario ensures, on the one hand, that the strongest generations join the reproductive population continuously and, on the other hand, that the number of new replenishments each year is relatively small as a percentage of the overall population, thus providing a certain stability to the total population. As a result, the annual variation in long-lived population age structure is relatively stable.

Short-lived populations consisting of fewer age groups have a simple structure and strong interannual variation in population size. The abundance or failure of a generation is quickly reflected in the population size. The population size decreases rapidly when marine environmental conditions are unfavorable and increases rapidly when they are favorable. On the other hand, the timing of initial sexual maturation is more consistent from generation to generation and thus strongly influences the variation in the size of the reproductive population.

(2) Type of Age Structure and Their Meanings

1. Type of Age Structure

Age structure includes single-age and multiage structures. Single-age structure refers to annual individuals, such as those of shrimp and most small and medium cephalopods; multiage structure is composed of multiple ages, such as those of most fish. The stability and variability in a multiage structure can be affected by conditions such as fishing intensity and generational abundance. For example, excessive fishing pressure has led to a decline in some traditional economic fisheries resources in the world's offshore areas with low age structures; in the East China Sea, in the late 1950s, the maximum age of striped bass was 6, and the proportion of those aged 1 and 2 years old was 77%; however, by the late 1970s, those aged 1 and 2 years old accounted for 98% of the population, and the maximum age was only 4 (Chen 2014; Chen and Liu 2017).

2. Relationship Between Age Structure and Changes in Population Size

The birth and mortality rates of a population have a strong influence on its age structure. The reproductive capacity of individuals in a population is often limited to certain age classes, and the magnitude of mortality rates varies with age. Thus, an analysis of age structure can predict the movement of a population's changing numbers.

In general, the age structure of a stock can be specified by the age structure of catches caught with representative gear in different fishing grounds and in different seasons. Usually, for an unexploited natural stock, changes in its age structure can be determined based on the following: (1) if the stock has a rapidly increasing resource, there are a large number of recruitment individuals, and the age composition of the catch is low; (2) if the stock is a stable resource, then its age structure is more evenly distributed and more constant; and (3) if the stock is declining and there are fewer recruitment individuals, then the proportion of older individuals in the catch is larger, and the age composition is higher.

In contrast, for an exploited stock, changes in the age structure of its catch can be determined based on the following: (1) if it is overexploited, then the age structure of the catch is significantly low; (2) if it is moderately exploited, then the age structure of the catch reflects its own typical characteristics; and (3) if it is underexploited, then the sequence of ages in the catch is long and the age composition is high. Thus, the age structure of the catch reflects the present status of the stock in terms of reproduction, replenishment, mortality, and abundance, which can be predictive of possible future scenarios. Therefore, compiling the age composition of catches over time is an important element of population dynamics studies.

In fisheries biology, frequency distribution charts or bar charts are usually used to represent the age composition of a population and its distribution (Fig. 2.1), which can visually reflect the age structure characteristics, dominant age composition, and dominant generations of a population. If the age composition data form a long time series, then it can clearly reflect the position of each generation in the population and their quantitative changes. As seen from Fig. 2.1, the maximum age of the stock is 9 years old; the 1991 generation of the stock, which dominated the catches in 1992, 1993, and 1994, was the dominant age group in the catches and became inferior in 1996, accounting for a small proportion. In addition, there were significant differences in the age compositions of the catches between 1990 and 1996. Therefore, the age structure of a stock is an important piece of basic information for studying changes in population size and preparing catch forecasts.

2.2.3 Concept of Individual (Length/Weight) Composition and Its Meaning

Individual composition usually includes body length composition and weight composition and refers to the proportion of the number of each body length group or weight group in a generation (or a stock). Usually, this variable is expressed as the individual (length and weight) composition of the catch.

In an undeveloped natural population, changes in the length composition and body weight composition of the population reflect changes in

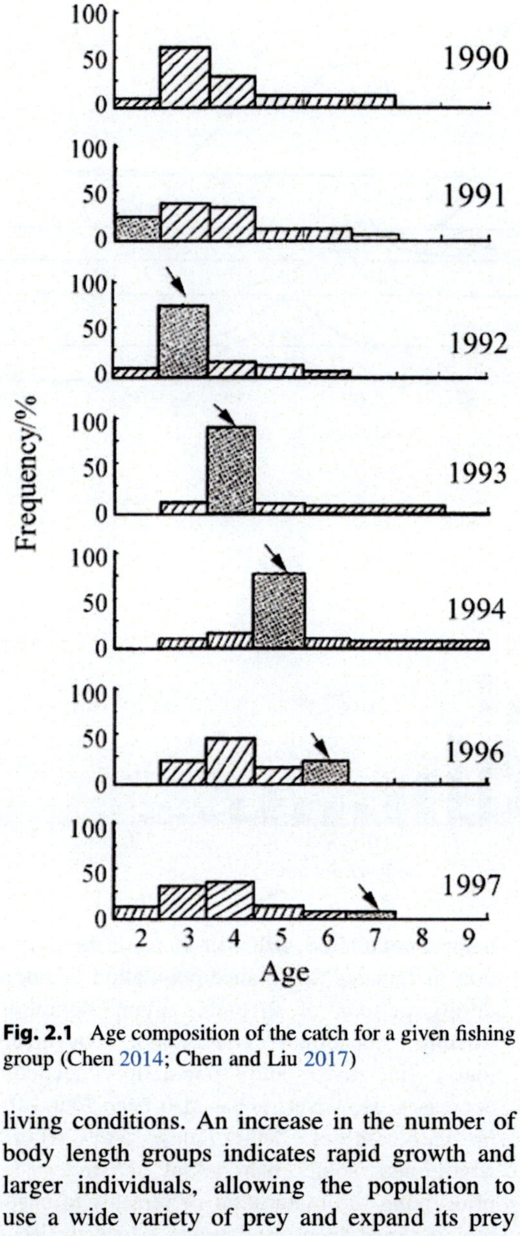

Fig. 2.1 Age composition of the catch for a given fishing group (Chen 2014; Chen and Liu 2017)

living conditions. An increase in the number of body length groups indicates rapid growth and larger individuals, allowing the population to use a wide variety of prey and expand its prey base, thus guaranteeing a more stable recruitment population; in contrast, if the number of body length groups is shortened, then the prey base is presumed to be unstable.

Because information on length composition and weight composition is easier to obtain than information on age composition and because percentage composition is quickly calculated and plotted as a frequency distribution or bar graph, length composition and weight composition have become the most fundamental elements of biological studies on fisheries resources (Fig. 2.2). These factors are especially important for species for which age identification is difficult and time-consuming or for which no age markers are available.

In addition, body length composition or weight composition distribution maps have been used in the delineation of populations and groups. Often, several groups with different ages or birth times can be distinguished using mixed distribution analysis of body length frequencies (Fig. 2.3), such as jumbo flying squid distributed off Peru, which is the largest individual species in the oceanic squid Ommastrephidae. As shown in Fig. 2.3, mantle lengths of male squids range from 203 to 736 mm, with dominant mantle lengths ranging from 230 to 440 mm, accounting for 80.0% of the total population, and the mean mantle length is 388.3 mm. Mantle lengths of female squids range from 205 to 805 mm, with dominant mantle lengths ranging from 260 to 440 mm, accounting for 74.4% of the total population, and the mean mantle length is 390.6 mm. Three peaks were observed for the mantle length composition of males, with mean values of 261 ± 21.5 mm, 381 ± 40.0 mm, and 496 ± 110.6 mm for the three groups, and three peaks were also observed for the mantle length composition of females, with mean values of 289 ± 32.2 mm, 406 ± 57.9 mm, and 634 ± 64.1 mm for the three groups (Chen et al. 2011).

In fisheries stock assessments, information on body length and weight composition can be converted into corresponding ages, and other biological parameters of a fishery, such as growth or mortality, can be derived. In fisheries science, the individual composition of a catch is also an important indicator for an analysis of a fishery. If the length composition of individuals in the catch is homogeneous, then it indicates the onset of the peak season; if the individuals of the species caught are extremely heterogeneous, then the season is coming to an end or is in an unstable fishing ground.

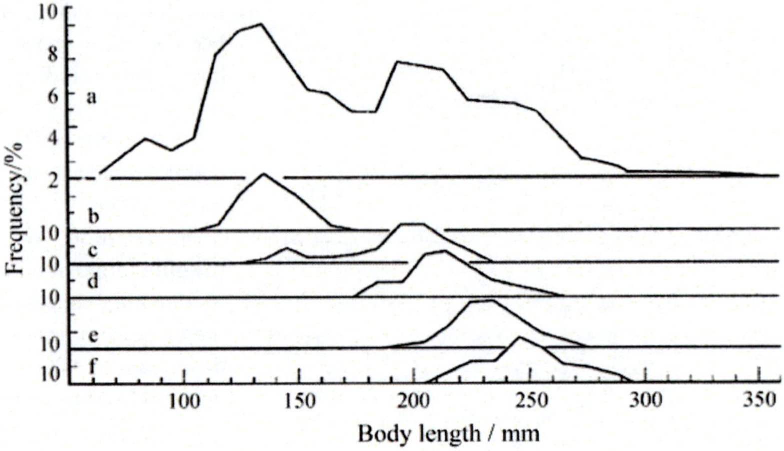

Fig. 2.2 Length composition distribution of a population and its body length distribution in groups aged 1–5 years old (Chen 2014; Chen and Liu 2017)
a is a composite distribution of length composition; *b*, *c*, *d*, *e*, and *f* are body length distributions for groups aged 1–5 years old

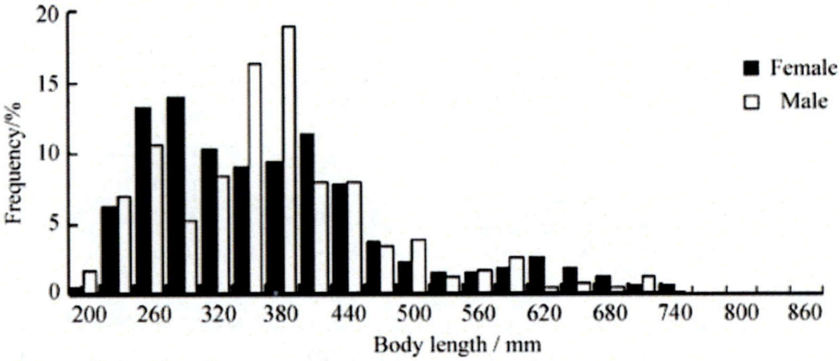

Fig. 2.3 Composition of mantle length of jumbo flying squid off Peru (Chen et al. 2011)

2.2.4 Sex Ratio, Sexual Maturity Composition, and Their Variation

(1) Sex Ratio Composition and Its Variation

Sex ratio composition is the ratio of the number of males to females in a population, usually expressed as the ratio of the number of males to females in a catch. Sex ratios in fish are usually regulated by altering metabolic processes through the following process: changes in food security → changes in material metabolic processes → changes in endocrine action → sex formation.

In catches, the sex ratio composition reflects the characteristics and changes in population structure, and such changes are a way of naturally regulating the population. In an unexploited natural population, fish will increase the proportion of females to enhance population fecundity during periods of favorable living conditions (mainly nutritional conditions); conversely, males increase, and population fecundity decreases. However, it has also been found that in the case of deteriorating prey security conditions, some fish adopt a strategy of prioritizing the maturation of females to ensure the continuation of the species, which is also an adaptation of that population to changing environmental conditions.

In addition, the sexes of individuals in some fish stocks switch under certain conditions at different developmental stages. For example, black bream are all males at the early age of sexual maturity and gradually change to females as they develop and grow, with females dominating the older age group. Groupers, on the other hand, are all females at early ages and become males only through sexual reversal at advanced ages, so the sex ratio of these species varies with age composition. There are also some fish species, such as the half-smooth sole, that have separate male and female habitats during the nonbreeding season, which also results in seasonal and geographic variation in sex ratio composition, but these traits are also an adaptive property of the population to changing environmental conditions.

In general, the sex ratio composition of marine fish stocks is most often approximately 1:1. The sex ratio composition is also related to fish growth, age, season, and other external factors, such as fishing.

(2) Sexual Maturity and Its Variation

Sexual maturity and its composition are important elements of population structure. Sexual maturity composition usually refers to the proportion of the number of each sexual maturity class in a generation (or a population). Usually, sexual maturity is expressed as the composition of individual sexual maturity levels in a catch. The division of sexual maturity varies among different categories, such as fish and cephalopods.

The age and duration of gonadal development and initial sexual maturation in fish vary from species to species. Within the same population, early and late sexual maturation times for individuals are related to changes in their growth rates and living environments, and gonadal maturity can reflect external influences on the population; e.g., if the water temperature is suitable and growth is good, then maturity will be faster. At the same time, sexual maturity, as an adaptation to population size regulation, can be reduced or increased by population size, which can lead to an earlier or later age of sexual maturity. In comparison to that of other species, the age of sexual maturity of traditional economic species such as small yellowtail and striped bass in China's coastal waters is significantly earlier, which fully reflects the reality of population decline.

The sizes of the individual fish at first sexual maturity are an important element of research when studying the biology of fisheries resources. The proportion of sexually mature individuals within different length groups and length group data were fitted to logistic curves using linear regression (Fig. 2.4) to derive the first sexually mature body lengths with the following equation:

$$\mathrm{Pi} = \frac{1}{1 + \mathrm{e}^{-(a+b*Li)}}$$

where Pi is the percentage of mature individuals in the group sample and Li is each body length group. Body length at first sexual maturity $(L_{50\%}) = -a/b$.

The composition of the sexual maturity levels of a population is usually divided into the recruitment part and the residual part. Knowing the composition of the complementary and residual parts not only provides timely information on changes in population structure but also is greatly important for studying and analyzing population dynamics. The recruitment part of the population refers to the part of the spawning population that reaches sexual maturity for the first time; the residual part refers to the part of the population who repeat sexual maturity.

In fisheries science, sex ratios comprise pairs of reproductive fish populations on spawning grounds, and these values can indirectly reflect the general trend in the fishery and its current state. For example, during the reproductive period, the number of males and females is approximately the same but varies slightly at various stages of the reproductive process. The pattern is as follows: at the beginning of reproduction, there are more males than females; at the peak of reproduction, the number of males is approximately the same as that of females; at the end of reproduction, there are fewer males than females. For spawning grounds, when sexually immature individuals are in the majority, the fishery has not reached the spawning stage, and the population is unstable; if gonads are in the majority, then the spawning stage is near, and the population is stable; and if spawned individuals are in the majority in the catch, then the spawning stage is nearing the end, and the population is less stable.

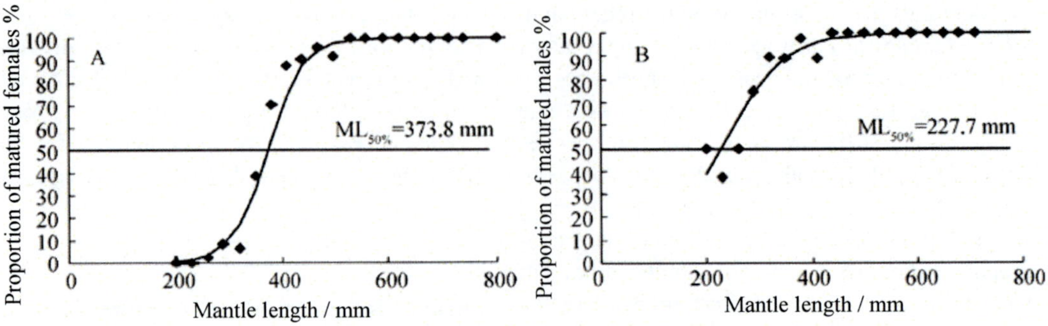

Fig. 2.4 Distribution of mantle length curves for first sexual maturity in jumbo flying squid off Peru (Ye and Chen 2007)

2.3 Methods of Stock Identification

2.3.1 Methodology for Stock Identification

The methods used to identify stocks are generally morphological, ecological, physiological, statistical, biochemical, and genetic. The first four methods are traditional, and with the development of computer technology, mathematical statistics, and artificial intelligence technology, new information and methodologies have become available in addition to these traditional methods. The last two methods (biochemical and genetic) have enriched the information available on population and group concepts to some extent and improved the accuracy of identification work, but these methods have also made population identification difficult due to the exchange of genes. Among these five methods, morphological methods are the most widely used. Population identification usually involves a combination of methods to reach a final conclusion.

Reproductive isolation and its degree of isolation are essential criteria for the delineation of species and populations, and reproductive isolation is also an important biological feature to prevent hybridization of organisms. Therefore, the material used for stock identification should generally be noted. (1) The sample must be a spawning population and sampled at the spawning ground so that it is reliably representative. The spawning population can provide information on biological characteristics such as morphology. (2) The sample should be fresh and intact, especially when using physiological and biochemical genetic methods, which require fresh sampling on site, and when using morphological methods, the scales and fins of the fish should be intact and undamaged to reduce the determination error. (3) The sample should be collected with sufficient representativeness and quantity, which are the prerequisites for conducting stock identification studies.

2.3.2 Morphological Approach

Morphological methods are traditional methods of identification. In taxonomy, the conventional procedure for identifying species is based on determining the morphological characteristics of an individual and its corresponding traits. The "characteristics" include both "qualitative" descriptions of the individual organism, such as the size of the fish, and "quantitative" measurements, such as segmentation and measurements of characteristic parameters. Because the morphological and genetic stability of species results in intermittent or significant differences in qualitative and quantitative characteristics among species, species identification is usually determined by obtaining a few individuals. In contrast, populations are groups of individuals within species, so they often show different degrees of continuity and trait variation in their characteristics and traits, which requires collecting samples of individuals from different spawning groups and obtaining a certain number; measuring and identifying their various segmentation characteristics, body size characteristics, and anatomical features; and then identifying populations according to the degree of variation in the characteristics of each sample. With the development of computer technology and image technology, geometric morphological methods have become an important method for population determination.

(1) Counting Characteristics

Counting characteristics mainly refer to the characteristics that are determined by counting and identifying various segmentation characteristics of fish before and after dissection and statistical analysis. The usual materials for determining segmental characteristics are vertebrae (number of vertebrae, number of carapace vertebrae, and number of caudal vertebrae), scales (number of lateral line scales, number of superior lateral line scales, number of inferior lateral line scales, and number of scutes), fin bars, number of pyloric caeca, number of gill rakers, number of gill cover bars, number of swim bladder branches, and number of scale phases and otolith rings.

1. Vertebrae

Vertebrae are divided into two parts: the trunk vertebrae and the caudal vertebrae. The caudal vertebrae are connected to the caudal rod bones at the end of the bones. The number of vertebrae is generally counted from the first vertebrae behind the skull to the caudal rods, but the caudal rods are not counted.

Among the morphological characteristics, the number of vertebrae is most commonly counted, and this indicator is used effectively to identify populations. In identifying populations, the number of trunk vertebrae and the number of caudal vertebrae are often counted separately. For fish with a very well-developed tail, such as the genus *Zoarces*, it is useful to count and identify the number of trunk vertebrae separately. However, in many cases, especially for herring, it is more appropriate to count only the total number of vertebrae. In short, how to count vertebrae and identify species vary from fish to fish. For example, when identifying the races of striped bass off China, some people use the number of trunk vertebrae and the number of posterior multimedial spine vertebrae, while others use the number of ventral vertebrae.

Patterns of variation in the number of vertebrae of the same species within similar species and ranges have been identified, with a general pattern of more vertebrae in the northern types than in the southern types and a gradual decrease in the number of vertebrae from north to south. This phenomenon also occurs in fjords and shallow waters, where the number of vertebrae is also lower than that in species in the outer sea.

In the past, the most common method used to obtain the number of vertebrae was dissection. However, when the number of specimens was small, especially in the case of type specimens and rare and valuable species, specimens had to be preserved intact and not dissected, so X-ray techniques were also applied to observe, count, and determine the number and morphology of vertebrae and other bones on film.

2. Scales

The number of scales in the lateral line is usually counted. The number of scales above and below the lateral line is counted from the base of the dorsal fin obliquely toward the lateral line, and the number from the base of the anal fin obliquely toward the lateral line is recorded separately. For example, for blue round scads, the number of prismatic scales on the lateral side of their bodies is counted, and for herring, the number of scutes on the venter is counted.

3. Fins

There are five types of fins on fish: dorsal, pectoral, ventral, anal, and caudal. The number of fins expresses the morphological characteristics of a population. The fins and spines of each fin should be counted and recorded separately. The type of fins to be used for stock identification should depend on the species. For example, in the analysis of geographic variation in morphological characteristics of the large yellow croaker (*Larimichthys crocea*), the comparison of fin morphological characteristics is usually given importance; it has also been suggested that the change in the number of anal fin of flounder counted in different years is the result of hydrological factors, with a change of 0.4 anal fins for a 1 °C change in mean temperature.

4. Pyloric Caeca

Many fish have many whisker-like, blind tubes called pyloric caecum that grow near the pylorus. The shape and number of pyloric caecum vary depending on the species of fish. When counting pyloric caeca, they should be determined by the total number of their bases. When counting, a dissecting needle should be used to examine them to avoid errors in counting.

5. Gill Rakers

Fish have gill rakers on the side of the gill arch facing the mouth, and generally, each gill arch has two rows of internal and external gill rakers, with the longest being the external gill rakers of the first gill arch. When using the number of gill rakers to identify a stock, the number of gill rakers on the first gill arch is generally used. Sometimes, depending on the species, the number of gill rakers in the upper and lower gill arches should be counted separately. The shape and configuration of the gill rakers vary depending on the diet of the fish. Generally, fish that feed on plankton have small gill rakers, while fish that feed on animal prey have a few coarse gill rakers. In addition, the number of gill rakers increases with age and growth, as in the case of Caspian herring. Therefore, when comparing differences in gill raker numbers between stocks or groups, it is advisable to do so by age group.

6. Swim Bladder Branches

The swim bladder branches of stonefish are often flanked by multiple pairs of lateral limbs branching in a dorsoventral direction, forming treelike branching on both sides of the swim bladder. Branching status is the basis for the identification of populations in rockfish. The complex branching of the swim bladder is a unique phenomenon in the family Sciaenidae, as in the case of each side of the swim bladder branch of the small yellow croaker, which usually divides the large branch into a dorsoventral branch and many smaller branches.

7. Scale Phases

Scales are generally measured by the distance from the first age to the nucleus (i.e., the radius of the first ring) and the number of rings in that distance, or the width of their resting bands is determined; in addition, the variation in the resting band coefficients of fish is also analyzed as a parameter for identifying populations. The resting band coefficient is the quotient obtained by dividing the distance from the nucleus of the scale to each resting band by the distance from the nucleus to the outer edge of the outermost resting band. Populations can also be identified on the basis of other scale characteristics. For example, to easily classify individual stocks, four ring types have been distinguished in Norwegian herring, namely, the northern type with distinct rings, the southern type with indistinct rings, the oceanic type, and the spawning type.

(2) Metric Characteristics

Metric characteristics are measured in terms of the length and height of the relevant parts of the fish, and the ratio between them is calculated and analyzed statistically, comparing the mean and the error of the mean. The body length ratios usually measured are full length/body length, body length/head length, body length/body height, head length/muzzle length, head length/eye diameter, and caudal peduncle length/caudal peduncle height. In addition, depending on possible differences in body type characteristics, maxillary length, postocular head length, eye diameter, dorsal fin base length, dorsal fin posterior length, anal length, pectoral fin length, and ventral fin length can also be measured, and various ratios can be calculated.

Counting traits and metric traits are the primary means of traditional stock identification methods and require extensive biological assay work. They are widely used because of the ease of sampling and measurement. For example, body shape metric characters (ratio of total length to anal length, ratio of anal length to head length, ratio of head length to snout length, and ratio of snout length to eye diameter) and morphological count characters of body segments (dorsal fin, pectoral fin, pyloric blind sac, trunk vertebrae, etc.) of *Trichiurus lepturus* samples from 12 sites in the coastal waters of China were generally divided into four groups: Yellow-Bohai Sea Group (63Y, 63P, 64P, and 64Y), East China Sea-Eastern Guangdong Group (63E1, 63E5, 64E2, and 64S1), Western Guangdong-Beibu Gulf Group (65 T, 64 T2, and 64S3), and Outer Beibu Gulf Group (64 T1) (Fig. 2.5).

Morphological traits are primarily controlled by genetic factors but are also influenced by environmental factors, such as water temperature, which has a strong influence on early development and variation in count traits such as fish vertebrae. The stability of traits, therefore, may show interannual variation and thus affect stock identification results. This approach requires extensive sample collection and biological assay work, followed by statistical analysis of morphological measurements to derive means and standard deviations to discern the degree of variation among the populations under study. Common statistical analysis methods include the following:

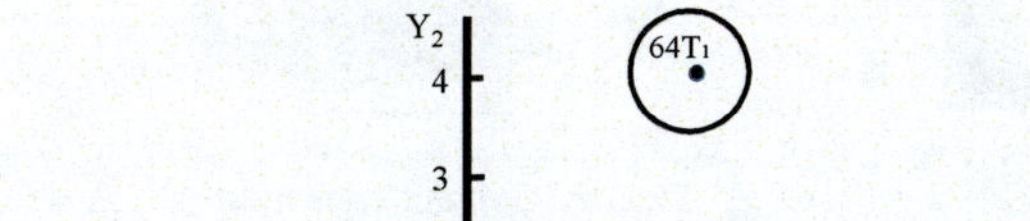

Fig. 2.5 Two-dimensional ranking of morphological characteristics of *Trichiurus lepturus* in 12 regions off China (Chen 2014)

1. Coefficient of difference (C.D.)

$$\text{C.D.} = \frac{M_1 - M_2}{S_1 + S_2} \tag{2.1}$$

where M_1 and M_2 denote the mean of the two population characteristic measures and S_1 and S_2 are the standard deviations of the two population characteristic measures.

To follow the rule for classifying 75% of subspecies, if a C.D. > 1.28 indicates variation at the subspecies level, then a C.D. < 1.28 is interpopulation variation.

2. Significance of difference in means (M_{diff})

$$M_{\text{diff}} = \frac{M_1 - M_2}{\sqrt{\frac{n_1}{n_2} m_2^2 + \frac{n_2}{n_1} m_1^2}} \tag{2.2}$$

where M_1 and M_2 are the means of the two population characteristic measures; m_1 and m_2 are the mean errors of the two population characteristic measures; and n_1 and n_2 are the sample sizes of the two population characteristics.

When $n_1 = n_2$ or a large number of samples, the denominator of the above equation can be simplified to $\sqrt{m_1^2 + m_2^2}$, calculated as a mean difference significance *t*-value test, and when $P < 0.05$, the difference is significant; if $P < 0.01$, then the difference is highly significant.

3. Discriminant function analysis

A multivariate discriminant function analysis can be applied to test for integrated differences in population characteristics, especially when the differences in single characteristics are not significant, to test for integrated differences between populations. The system of linear equations is the following:

$$\begin{aligned} \lambda_1 s_{11} + \lambda_2 s_{12} + \ldots + \lambda_k s_{1k} &= d_1 \\ \lambda_1 s_{21} + \lambda_2 s_{22} + \ldots + \lambda_k s_{2k} &= d_2. \\ &\ldots \\ \lambda_1 s_{k1} + \lambda_2 s_{k2} + \ldots + \lambda_k s_{kk} &= d_k \end{aligned} \tag{2.3}$$

From the above equations, the judgment coefficients $\lambda_1, \lambda_2 \ldots, \lambda_k$ can be solved and where d_i denotes the off-mean difference of i population characteristics; S_{ij} denotes the sum of the covariances of i and j population characteristics; k is the population characteristic term; $i, j = 1, 2..., k$; and the judgment function is $D = \lambda_1 d_1 + \lambda_2 d_2 + \ldots + \lambda_k d_k$.

The test of significance of the difference is

$$F = \frac{n_1 \times n_2}{n_1 + n_2} \times \frac{n_1 + n_2 + k - 1}{k} \times D \quad (2.4)$$

where n_1 and n_2 are the number of two samples.

According to the F-value test, the difference is significant when $F > F0_{,05}$ or $F0_{,01}$. In fact, the method is a combined evaluation of each indicator, calculating the total variability in each eigenvalue, $\lambda_1, \lambda_2, \ldots, \lambda_k$ corresponding to the weights.

With the development of mathematics and computers, there are an increasing number of statistical testing methods, such as the Fourier method, landmark point method, mean clustering, fuzzy analysis, gray clustering, and spatial distance analysis.

(3) Geometric Morphological Approach

Some of the more commonly used geometric morphometric methods are Fourier analysis and landmark point methods, which have been widely used in stock identification and fish morphology.

1. Fourier Analysis

The first geometric morphometric method used was the external feature method, also known as the external contour method. The method consists of selecting a certain number of sample points on the external contour of an object according to the principle of homology; coding the external contour after binarization using software; decomposing the contour curve after polarizing the coordinates into functions conforming to sine and cosine, thus deriving the Fourier coefficients; and finally finding the morphological differences by subsequent statistical methods (Fig. 2.6). The method was created by the French mathematician Joseph Fourier. The method is mainly based on elliptical Fourier analysis, and it has been widely applied to the study of otolith species identification because of the near-elliptical character of their otoliths.

The external contour of an otolith is not a smooth curve and does not meet the definition of a "curve," which does not result in errors in conventional measurements. In Fourier analysis, the focus is on the representation of the external contour, which in practice needs to be reconstructed as a mathematical curve function with wavelet transform analysis and curvature scale space analysis.

Fig. 2.6 Steps of the profile analysis method (Chen et al. 2017)

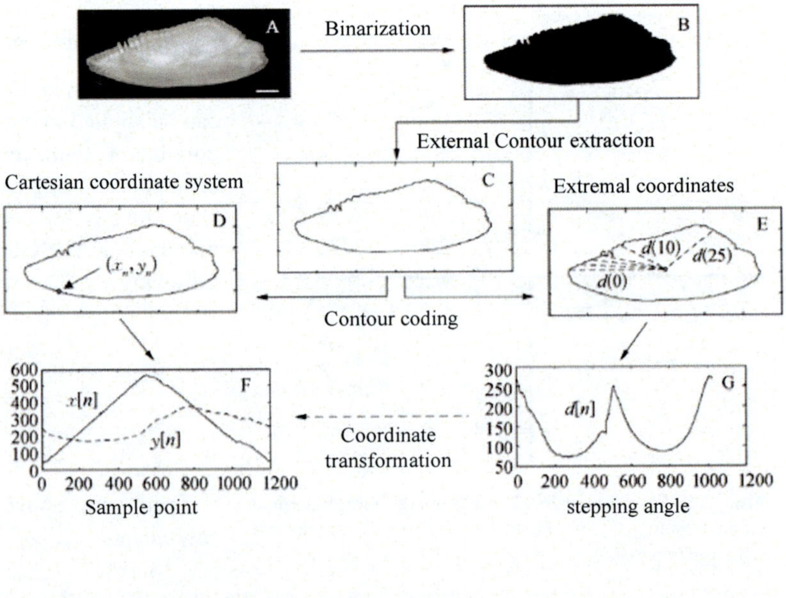

2. Landmark Point Analysis Method

The Fourier analysis method focuses on the selection of the overall external contour, and there is no uniform rule for the selection of the external contour of different objects; thus, there is a certain error in describing the external morphology. Therefore, the current description of morphology is mostly focused on the landmark point method. This method is based on the shape statistics method of Cartesian landmarks by obtaining a two-dimensional image of the object and transforming it into x and y coordinate point data. Landmark points follow certain principles in selection: (1) Type I landmark points mainly refer to the intersection points between different tissues, such as the connection point between bones and muscles and the connection point between fish body and fins. (2) Type II landmark points refer to the depression or projection points between tissues, such as the protrusion of bones and the gap of otoliths, or other points in the tissues that are in a prominent position and can be clearly identified and analyzed. (3) Type III landmark points refer to the most valuable points between tissues, such as the longest point and widest point (Fig. 2.7).

When landmark points are selected, there will always be errors due to the size, location, and orientation of the sample. If left untreated, then these factors can have an impact on subsequent analyses, so the overprinting method is needed to remove the interference. The least squares criterion is the most widely used overprinting method, and the overprinting effect is achieved by translating and rotating between samples. The morphological differences are then analyzed by thin-plate spline analysis (thin-plate spline), which uses deformation and deconvolution to make the coordinate values of multiple samples correspond and to draw a deformation grid. This method introduces the distortion energy matrix (bending energy matrix) into the geometric analysis, and the differences are compared by various statistical methods, such as local distortion (particle warp) and relative warp (relative warp) (Fig. 2.8). This method introduces statistical methods such as energy matrices, principal component analysis, and eigenvalue analysis and is now widely used in fish species and population identification.

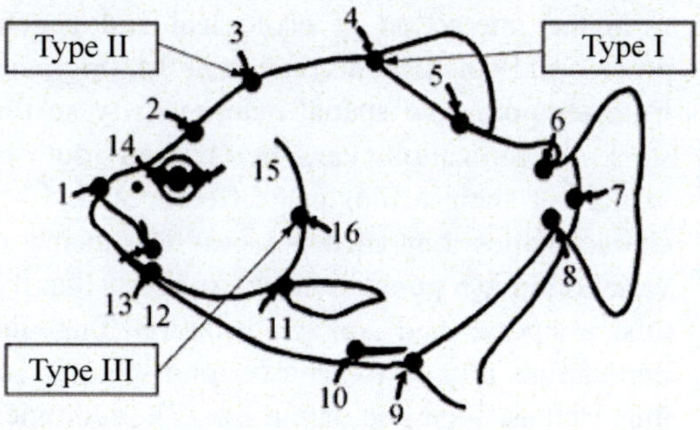

Fig. 2.7 Schematic diagram of the three types of landmark points (Chen et al. 2017)

3. Software Commonly Used for Geometric Morphometry

Geometric morphometry is generally based on two-dimensional images for analysis, so software is needed to convert the images into data. At present, the commonly used software is comprehensive analysis software such as MorphJ (Klingenberg 2011) and PAST (Hammer et al. 2001) and series software such as TPS (Rohlf 2013), IMP (Sheets 2013) and SHAPE (Iwata and Ukai 2002). Comprehensive software can individually complete the analysis steps from sample point acquisition to distortion analysis, while serial software consists of multiple programs that undertake different analysis steps to complete the analysis process together.

Professor F. James Rohlf of the Department of Ecology and Evolution at New York University at Stony Brook began his research on geometric morphometry in the late 1980s and developed the TPS series of software. The software package combines image digitization, overprinting, deformation grid analysis, and simple statistical analysis, such as principal component analysis and regression analysis, to perform correlation analysis of geometric morphometry for 2D images. The software package mainly consists of tpsUtil, tpsDig, tpsPLS, tpsRegr, tpsRelw, and other programs. Among them, tpsDig is the core element in the package, which mainly reads the

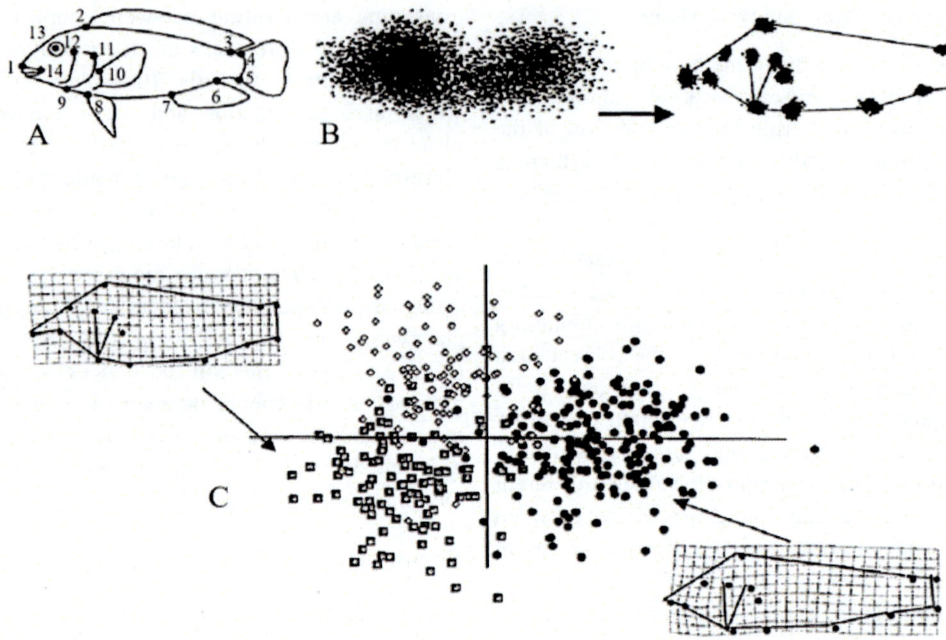

Fig. 2.8 Operational steps of the landmark point method (Chen et al. 2017) *A*. Quantification of unprocessed data (data taken from Cichliformes). *B*. Removal of the effect of nonmorphological variation (412 samples compared before and after Generalized Procrustes Analysis processing) *C*. Statistical analysis (Canonical Variance Analysis) and graphical representation of the results (rasterized deformation of the left-hand side *Spathodus erythrodon* and the right-hand side *Eretmodus cyanostictus*)

outline or custom landmark points of the study object in the image and coordinates them, playing a key role in digitizing image content (Fig. 2.9), and then, the analysis is completed using a variety of other analysis programs in combination.

TPS series software is easy to operate, has a user-friendly interface, and can obtain results relatively quickly. TPS software was developed in the 1990s, and many programs need to be run on earlier platforms (some programs can only be run in the DOS environment). The core part of tpsDig is liked by many researchers for its stability and simplicity of operation, and it has been used in various research analyses.

Hiroyoshi Iwata of the University of Tokyo has been working on plant morphology for a long time and has developed SHAPE software, which is mainly used for elliptical Fourier analysis in biomorphological research. The "Chain coder" program is for Grayscale, Binarize Image, and Chain coding; "Chc2Nef" is used to create Fourier harmonics; "PrinComp" is used for principal component analysis; and "Chcviwer" or "Nefviwer" can be used to observe the effect of recording changes in the edges of a target (Fig. 2.10).

2.3.3 Ecological Approach

In the oceans, the discrete nature of fish populations is a dynamic characteristic that arises from the interaction of ecological and genetic processes. Ecological dispersion in fish originates from temporal and spatial heterogeneity so that stock identification can be carried out by exploiting their ecological differences and the characteristics they each possess. The theory of variation in fish population size suggests that it is this temporal and spatial isolation that fish populations rely on to achieve food security and thus increase their population size. The ecological approach is therefore one of the most important methods of stock identification.

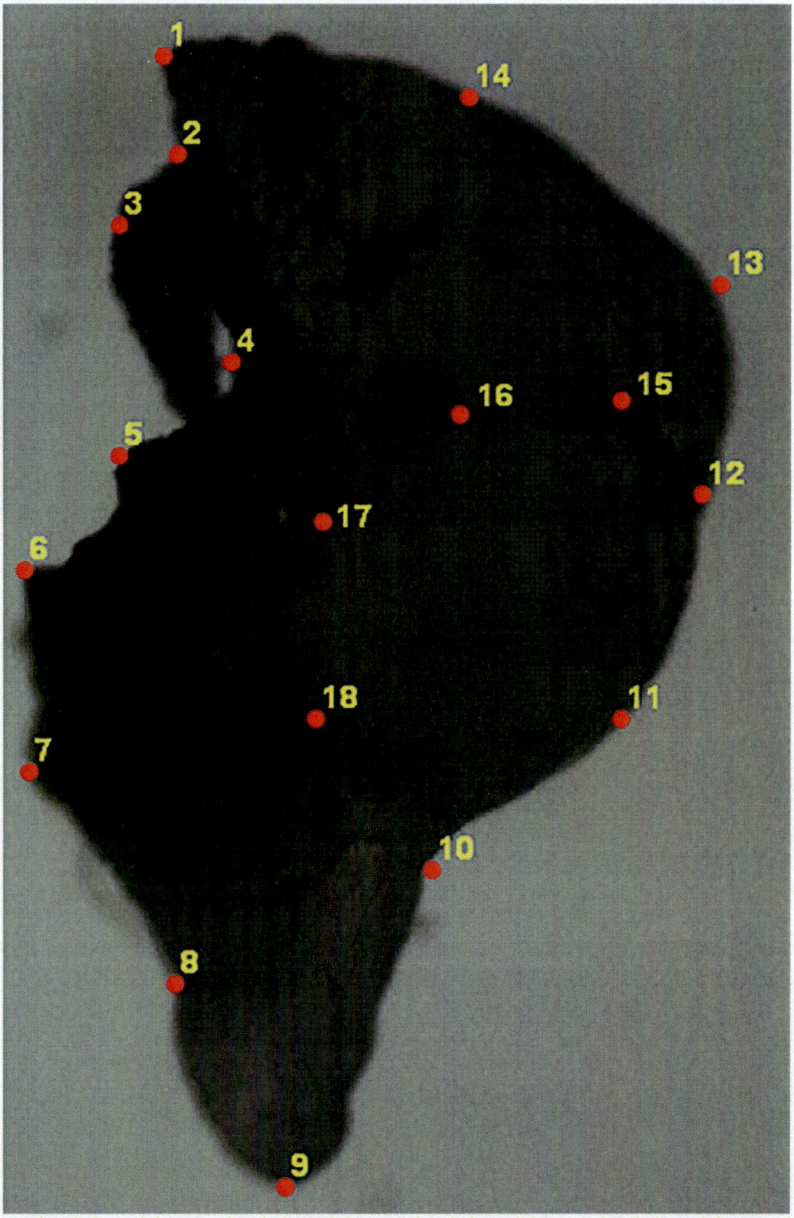

Fig. 2.9 Cephalopod statolith images and their landmark points (made by tpsDig V2.16) (Chen et al. 2017)

The ecological approach involves studying and comparing the life histories of populations and their parameters under different ecological conditions. The main indicators are as follows: (1) reproductive indicators, reproductive period, egg carrying capacity, fecundity, spawning capacity, etc.; (2) growth indicators, length and weight, growth rate, fecundity, etc.; for example, the growth rate change of the same age group can be used as an important basis for small yellow croaker stock division; (3) age indicators, life span, age composition, age of sexual maturity, etc.; for example, geographical variation in the relative growth of otoliths and body length can be used to divide the *Trichiurus lepturus* stock off China; (4) migration distribution, differences in

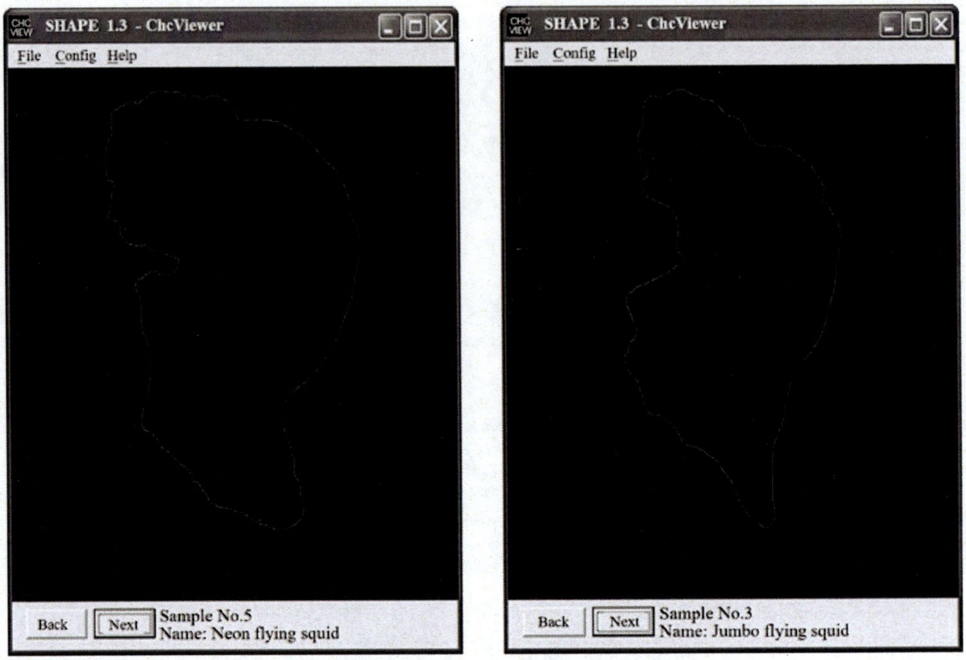

Fig. 2.10 Schematic outline of different oceanic squid statoliths (Chen et al. 2017) (**a**) is *Ommastrephes bartramii*; (**b**) is *Dosidicus gigas*

the migration routes of the stock as a basis for stock division; for example, marker release methods can be used to study the migratory distribution of shrimp in the Yellow and Bohai Seas, which in turn confirms that shrimp in the Bohai Sea and the north-central Yellow Sea belong to the same population; (5) feeding indicators, feeding species, feeding frequency, etc.; (6) the rhythm of population size changes; (7) parasites, the species of parasites, etc.; and (8) trace elements, trace elements in hard tissues and their composition; for example, these elements have become one of the main materials for the study of fish populations, especially some estuarine species. The following indicators are commonly used to identify populations:

1. Migratory distributions. The most direct method is marker release. It is also used to determine the timing and routes of the various migrations of a stock and the distribution range of overwintering grounds, spawning grounds, and prey grounds and to investigate the differences in the migratory distribution of juvenile and adult fish through systematic surveys of resource fisheries. For example, based on a large number of systematic fisheries resource surveys, the adaptation of *Trichiurus lepturus* to the natural regulation of marine environmental conditions was studied, and the distribution of *Trichiurus lepturus* from south to north was divided into three stocks; based on the recapture results of Japanese amberjack marker release and the migration range in the western Pacific, Japanese amberjack was divided into two different stocks, northern and southern, using the tidal headland north of 33°N as the boundary.
2. Comparisons of growth, reproductive habits, and age composition. Different populations have different growth conditions due to different living environments so that different populations can be identified based on their growth differences. For example, according to the comparison of the relative growth of body length and pure body weight and the regression parameters of pure body weight

and body length of three geographic populations and eight reproductive groups of large yellow croaker in the Chinese coastal waters, it was found that there were differences in the relative growth of both pure body weight and body length among the subspecies groups, and the regression parameters of pure body weight and body length also differed to a greater or lesser extent.

Comparisons of reproductive habits included age at maturity, spawning time, egg mass, and egg diameter of each group. For example, the age of first sexual maturity and the age of mass sexual maturity of large yellow croaker increase from south to north; for different groups of large yellow croaker of the same race, there are different degrees of differences in spawning time, such as spring spawning and autumn spawning.

In addition to the several ecological habits mentioned above for stock identification, ecological habits can also be used as a reference for stock identification based on studying differences in the sensory physiological functions of a species in response to external conditions, as well as to differences in the external environmental conditions under which a species lives, such as the size of spawning grounds and water depth.

3. Parasite markers. Populations inhabiting different waters often have their own inherent parasite compartments. Therefore, certain biological indicators can be found on some fish bodies to distinguish them, such as *Coilia* in the Yangtze River, which has both within river and river-sea migratory groups. When these groups mix, they can be distinguished by the presence or absence of marine parasitic crustaceans on their bodies. Similarly, the different river systems where this fish are born can be distinguished from those of salmon in their marine habitat by the type of parasitic organisms in the fish.
4. Hard tissue trace elements. During the exchange of substances between fish, cephalopods, and other animals and the external environment, chemical elements from the environment enter the body through respiration, ingestion, etc., and are then deposited in hard tissues such as otoliths after a series of metabolic and circulatory processes that form them into endolymphatic crystals. These elements are deposited in otoliths in very small amounts after decreasing during transport through the body, and they are called trace elements. Trace elements can be divided into small elements (minor elements, concentration $> 100 \times 10^{-6}$, e.g., Na, Sr, K, S, N, Cl, and P) and trace elements (concentration $< 100 \times 10^{-6}$, e.g., Mg, Cu, Pb, Hg, Mn, Fe, and Zn) according to their content. Due to the noncellular and metabolically inert nature of hard tissues, the chemical elements deposited in hard tissues such as otoliths in the aquatic environment are essentially permanent as the fish and their otoliths grow in parallel. Hard tissues such as otoliths record the characteristics of the aquatic environment in which they live throughout their life cycle, and changes in the aquatic environment lead to changes in trace elements in hard tissues. An analysis of information related to the surrounding water environment and trace elements in hard tissues can not only effectively classify groups but also play an important role in the analysis of the life history of fish, such as migrating, breeding, and spawning, as well as in the reconstruction of habitats and their characteristics such as temperature, salinity, and food.

The main methods currently used for trace element analysis of hard tissues such as otoliths are the following: inductively coupled plasma-mass spectrometry (ICP-MS), laser ablation-inductively coupled plasma mass spectrometry (LA-ICPMS), proton-induced X-ray emission (PIXE), synchrotron X-ray fluorescence spectroscopy (S-XRF), electron probe microanalysis-electron microprobe analysis (EPMA), and nanosecondary ion mass spectrometry (nanoS IMS), atomic absorption spectrometry (AAS), inductively coupled plasma with optical spectrum analysis (ICP-OES), and proton backscattering (BS).

5. Oxygen and carbon isotope analysis methods. Oxygen and carbon isotope analysis methods for fish stock identification were developed on the basis of microstructural studies of fish otoliths. Compared with genetic analysis methods, isotopic signatures of fish otoliths have two significant advantages: first, the ring-band structure of fish otoliths provides an ideal time series for separating different growth stages of fish (especially marine fish); second, these signatures provide information corresponding to the fish living environment based on the formation mechanism of fish otoliths, making it possible to reconstruct the growth history of fish. In particular, it should be noted that the $\delta^{18}O$ of fish otoliths reflects the water status of the fish habitat, while $\delta^{13}C$ reflects the food status of the fish, and the combination of the two isotopic components is a useful tool for identifying fish populations and groups.

For example, Atlantic salmon (*Salmo salar*) in the Gulf of Maine in the northeastern United States is a protected species. Proper identification of farmed versus wild salmon from different hatcheries is extremely important for resource conservation. The study collected 40–50 Atlantic salmon otolith samples from three farmed hatcheries and two wild hatcheries for a $\delta^{18}O$ and $\delta^{13}C$ compositional analysis and compared the raw analytical values or means of the two sets of data to determine whether juvenile fish from the five hatcheries could be separated. The otolith stable isotope raw data and mean values correctly and effectively determined Atlantic salmon from different hatcheries (Fig. 2.11). Therefore, discrimination criteria based on otolith isotopic compositional characteristics have the potential to be used in practice to support resource management of Atlantic salmon in the Gulf of Maine.

2.3.4 Molecular Biology Approach

Since the 1980s, with the continuous development of molecular biotechnology, genetic markers of molecular ecology have been developed from the protein level to the deoxyribonucleic acid (DNA) level, and a wide variety of DNA marker technologies have emerged. In particular, expressed sequence tag

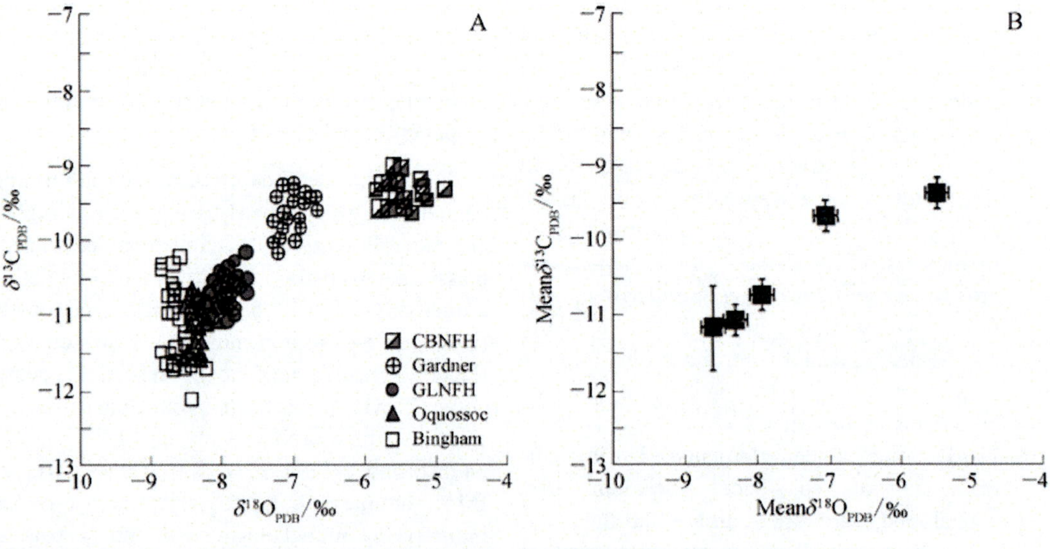

Fig. 2.11 Results of stable isotope composition analysis of Atlantic salmon otoliths (Thomas et al. 2008) (**a**) Raw otolith isotopic composition data; (**b**) mean otolith isotopic data

(EST) technology has started to produce ribonucleic acid (RNA) level markers. At the same time, with the continuous updates and improvements in molecular marker technology tools, new levels of marker technology (such as the perfection of RNA level) will continue to emerge to promote the application of molecular biology in population identification. Currently, molecular biology methods include serum agglutination reaction, isozyme electrophoresis, restriction fragment length polymorphism (RFLP), random amplified polymorphic DNA (RAPD), and amplified fragment length polymorphism (AFLP), simple sequence repeats (SSRs), intersimple sequence repeat (ISSR) markers, mitochondrial DNA markers, single-strand conformation polymorphism (SSCP), single nucleotide polymorphism (SNP) markers, EST, and other technologies.

(1) Protein-Level Labeling

Currently, the main methods used for protein-level labeling of molecular population geography are serum agglutination reactions and isozyme electrophoresis; allelic enzyme labeling is less common in the study of actual populations.

1. Serum Agglutination Reaction Method

The serum agglutination reaction method was developed on the principle that many organisms have a protective reaction when an infectious agent or a particular heterogeneous protein (antigen) invades the organism from a pathway other than the intestine, where its plasma proteins play an important protective role. The protective reaction of the organism is manifested in the formation of special protein bodies called antibodies, which enter the plasma and make the infectious agent harmless only after meeting with various antigens. The serum agglutination reaction method is based on this principle for the identification of fish stocks.

This process is conducted by taking fish proteins as antigens, injecting them into rabbits or other test animals, and causing them to produce antibodies, and after a certain period of time, their blood is drawn to produce a serum called an antiserum. Since this serum contains antibodies, it produces a cloudy precipitate when the fish protein that was originally used as the antigen is dropped on top of it. This reaction is known as the serum agglutination reaction. The corresponding protein of a closely related fish reacts to this antiserum, and the more distantly related it is, the less precipitate is produced. This precipitation is used as an indicator for the identification of fish populations. For example, this method has been successful in identifying coastal striped bass populations in southeastern China.

2. Isozyme Electrophoresis Method

Genetic methods are mainly used to determine the substances analyzed by electrophoresis techniques, such as fish protein molecules that are charged and mobile in a certain buffer and different protein molecules in different loci of the population, which show different mobility, as a criterion to discriminate the population. A more accurate identification approach is to perform isozyme electrophoresis analysis and electrophoresis to obtain the phenotype and its frequency and thus calculate the allele frequency and genetic distance. Differences in heritability between populations mainly occur in gene frequencies, while differences between individuals of the same population generally lie in allelic differences.

For example, Chinese scholars studied the biochemical genetic structure of and variation in silver carp and bighead carp populations in the Yangtze, Pearl, and Heilongjiang river systems using plate electrophoresis and polypropylene discretionary gel electrophoresis in the mid-1980s. The study showed that there was obvious biochemical genetic variation among the populations of the same species in the different water systems; for example, the proportions of polymorphic loci were 13.3%, 26.7%, and 13.3% in the Yangtze, Pearl, and Heilongjiang river silver carp populations, respectively; the average heterozygosity was 0.0493, 0.0484, and 0.0511, and their codon differences were 0.0506, 0.0496, and 0.0525, respectively. At the same time, the proportion of polymorphic loci in the southern population tended to be higher than that in the northern population. The genetic similarity and genetic distance of the Yangtze River silver carp-Pearl River silver carp, Yangtze River silver carp-Heilongjiang silver carp, and Pearl River silver carp-Heilongjiang silver carp were 0.9957, 0.0043, and 0.9955 and 0.0045, 0.9696, and 0.0304, respectively. The genetic differences were small, while the genetic differences between the Heilongjiang population and the above two populations were large (Fig. 2.12).

(2) Molecular Marker Technology at the DNA Level

DNA molecular marker technology can overcome the disadvantages of a low number of protein-level markers and susceptibility to environmental influences. With the increasing level of sequencing and sequencing accuracy, DNA molecular marker technology is widely used to identify the genetic structure of populations.

1. Restriction Fragment Length Polymorphism (RFLP)

Restriction fragment length polymorphism was originally proposed to identify DNA polymorphisms. This method mainly uses restriction endonucleases to enzymatically cleave the genomic DNA of individuals, transfer the enzymatically cleaved DNA to hybridization membranes by electrophoresis and Southern blotting, and select certain probes with which to hybridize, thus showing the difference in the lengths of enzymatic fragments containing homologous sequences with the probes. This difference is essentially caused by mutations in single bases or insertions, transfers and inversions of structures, etc., compounding the Mendelian inheritance of codominant marker techniques.

For example, Shan et al. (2006) studied the population genetic structure of natural populations from Ruichang of Jiangxi province, Changsha of Hunan province, and a captive-bred bighead carp population from Ninghe of Tianjin, using RFLP techniques and found that there were significant genetic differences between the Changsha population and the Ruichang and Changsha populations (Fig. 2.13).

Although RFLP is technically feasible in terms of studying population structure, it requires high sample purity, a relatively large number of samples, limited polymorphic information content of single digests, and too great a dependence on the type and number of restriction endonucleases. RFLP analysis is technically multistep and requires a large amount of work, usually requiring a large number of samples to analyze population structure, thus increasing the cost of the experiment.

2. Random Amplified Polymorphic DNA (RAPD)

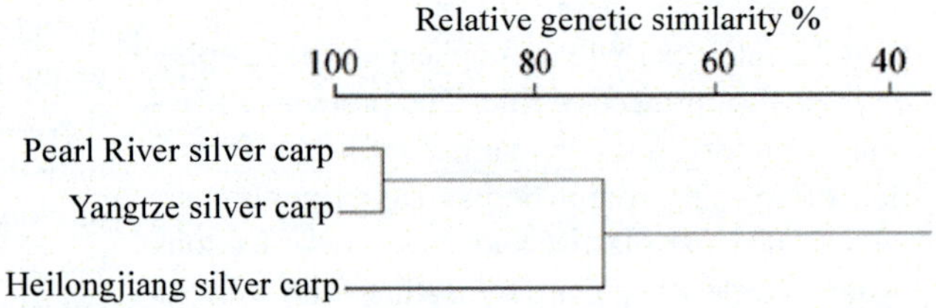

Fig. 2.12 Cluster analysis of the genetic similarity among three silver carp populations (Li et al. 1986)

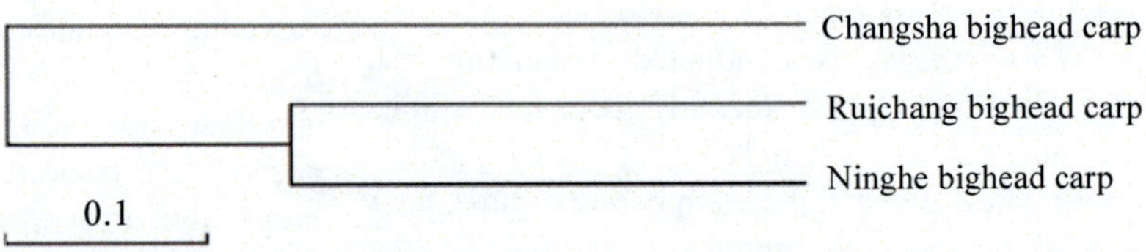

Fig. 2.13 UPGMA clustering of Rogers genetic distances for three populations of bighead carp (Shan et al. 2006)

Random amplified polymorphic DNA is a molecular marker technique. This technique builds on PCR and uses a series of single-stranded random primers with approximately ten bases to amplify through PCR all of a genome's DNA. When there are long or short inverted repeats spaced apart in the genome, PCR amplification of the gene by a single short random sequence of primers is able to produce a repeatable amplified fragment as long as the spacing of the two primer binding sites, which are complementary in the opposite direction, meets the conditions for PCR amplification. Changes in the DNA sequence of the primer binding site and deletions, insertions, or substitutions of DNA bases between the two amplification sites can lead to differences in the number and length of amplified fragments, resulting in detectable polymorphisms.

For example, Zeng et al. (2012) studied the genetic diversity of *Trachidermus fasciatus* populations in four different rivers, the Fuchun, Yellow, Luan, and Yalu rivers, using RAPD techniques, and the genetic distance among the four populations ranged from 0.0082 to 0.0246. UPMGA cluster analysis showed that the Yalu, Yellow, and Fuchun river populations clustered into one group, and the Luan River population was a separate group (Fig. 2.14).

The RAPD technique is a highly scientific method to determine population structure; however, the plots often have certain weak bands, and their reproducibility is poor. Moreover, the primer length, sequence, number of primers, amplification reaction system, and amplification conditions were not standardized in this technique, which made the test results vary.

3. Amplified Fragment Length Polymorphism (AFLP)

AFLP is a DNA molecular marker technique. This technique is based on selective PCR amplification of genomic restriction fragments. As genomic DNA varies in size from species to species, single nucleotide mutations or insertions and deletions result in increased or decreased restriction endonuclease cleavage sites. The type, number, and order of the selective bases determine the specificity of the amplified fragment. The amplification products are separated by radioisotope labeling and polyacrylamide gel electrophoresis, and then, the polymorphism of the products is detected by the presence or absence of DNA fingerprints on the gel.

For example, Ying (2011) studied the genetic diversity of Japanese sardinella populations (Rizhao population, Qingdao population, Zhoushan population, Aichi population, and Kagawa population) using five pairs of AFLP primers and showed that the genetic differentiation index Fst ranged from −0.0027 to 0.0827. Overall, the Fst between Chinese populations was small, and the P value was greater than 0.05, indicating that there was no significant genetic differentiation between Chinese populations. The Fst between the Chinese and Japanese populations was significant, indicating that the Chinese and Japanese populations belonged to two different populations. uPGMA trees indicated (Fig. 2.15) that the Japanese and Chinese populations were two different populations.

The advantages of AFLP marker technology, such as better reproducibility and more marker sites, over other technologies result in it being widely used to study the genetic diversity of marine organisms. However, the high cost of the kits used during AFLP experiments limits the promotion of this technology; the operation requires the use of isotope markers, which are relatively expensive; and the sample DNA

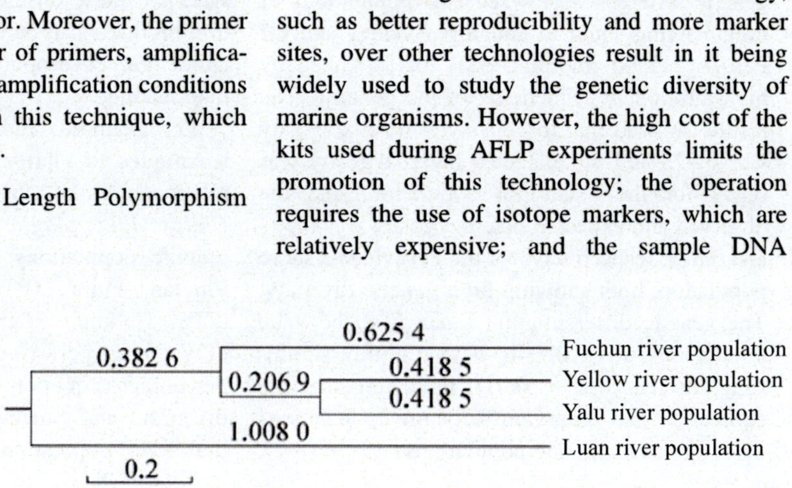

Fig. 2.14 UPGMA clustering analysis among four *Trachidermus fasciatus* populations based on genetic distance (Zeng et al. 2012)

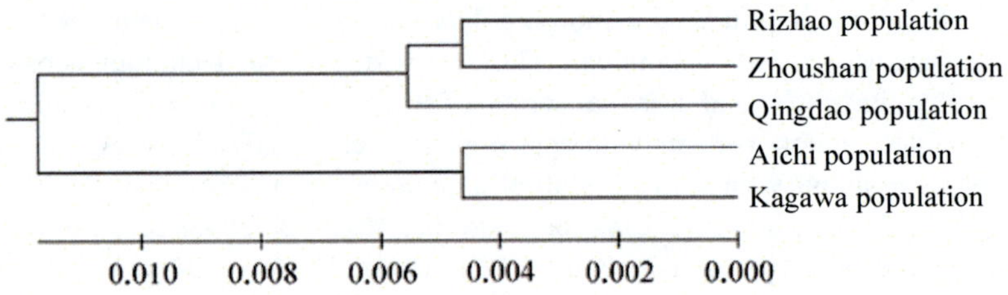

Fig. 2.15 UPGMA tree constructed based on genetic distances between populations of Japanese sardinella (Ying 2011)

integrity within the study fragment must be high quality.

4. Microsatellite Simple Sequence Repeats (SSRs)

Microsatellite DNA sequences are stringlike simple repeats widely distributed in eukaryotic genomes. Each simple repeat unit is between 2 and 10 bp in length, and common microsatellites, such as TGTG ...TG, notated as (TG)*n* or AATAAT...AAT, denoted as (AAT)*n*, are known as simple sequence repeats. Different numbers of sequences are arranged in tandem repeats, thus exhibiting length polymorphism. On both sides of each SSR are generally relatively conserved single copy sequences, whereby primers can be designed. The amplified SSR sequences are subjected to polyacrylamide gel electrophoresis and silver staining, and the polymorphisms of different individuals on a particular SSR are read by comparing the relative migration distances of the bands.

For example, Liu et al. (2015a, b) analyzed the genetic differences between two populations of jumbo flying squid in equatorial waters and off Peru by 12 polymorphic SSR loci obtained by high-throughput sequencing of the genome. The results showed that the observed heterozygosity was 0.725 and the expected heterozygosity was 0.897 for the equatorial population, and the observed and expected heterozygosity was 0.697 and 0.874, respectively, for the Peruvian offshore population, both showing high genetic diversity. The genetic differentiation index (Fst) between populations was 0.02046, with a highly significant difference ($P < 0.01$), indicating that the equatorial and Peruvian offshore populations belong to two separate populations.

A SSR is widely distributed in the genome, and its polymorphism is good. The SSR is designed to detect the primary structure of DNA, and this method is effective for both heterozygotes and pure heterozygotes, has good reproducibility, and can be automated. The detection of one locus can be completed within 24 h, or multiple loci can be detected simultaneously.

5. Intersimple Sequence Repeat (ISSR) Markers

An ISSR is a novel molecular marker technique, mainly used for detecting DNA sequence differences between SSRs, and is a further extension of the SSR marker technique. An ISSR uses anchored microsatellite DNA as primers and adds 2–4 random nucleotides to the 3′ end or 5′ end of the SSR sequence. During PCR amplification, the anchoring of the primers can cause annealing of specific sites, causing PCR amplification of repetitive DNA fragments that are not too widely spaced in complementary intervals to each other with the anchored primers. For this reason, the products between the amplified SSR gaps vary in size, and these differently sized PCR amplification products can be distinguished by polyacrylamide gel electrophoresis, silver staining, and map reading.

For example, Pan et al. (2018) used ISSR techniques to clarify the genetic diversity and relationships of the Chinese population of *Chiloscyllium plagiosum* in five indigenous geographic populations of Xiamen, Fujian (XM); Pingtan, Fujian (PT); Zhanjiang, Guangdong (ZJ); Haikou, Hainan (HK); and Taiwan (TW) and performed genetic analysis. The percentages of polymorphic loci, Nei's gene diversity, and Shannon's information index of the five populations were 38.27%–58.02%,

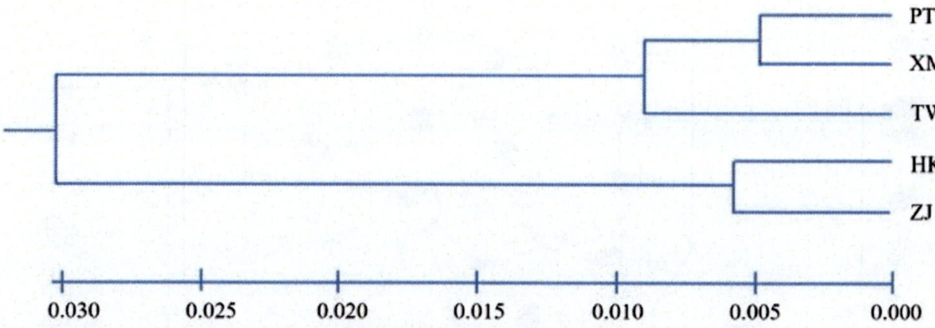

Fig. 2.16 UPGMA clustering map of five populations of *Chiloscyllium plagiosum* (Pan et al. 2018)
XM Xiamen population, *PT* Pingtan population, *ZJ* Zhanjiang population, *HK* Haikou population, *TW* Taiwan population

0.1353–0.2155 and 0.2032–0.3193, respectively. uPGMA clustering analysis showed that the PT population clustered first with the XM population, then with the TW population, and finally with the ZJ and HK populations forming branching clusters (Fig. 2.16), i.e., a pattern of gene exchange associated with geographical proximity.

The ISSR approach overcomes the shortcomings of SSRs, which require DNA library construction while maintaining no weaker polymorphism than the SSR markers, making the ISSR approach more reliable, simple to operate, and relatively inexpensive, with less DNA used and high safety performance. For this reason, ISSR marker technology has been rapidly and widely used in population genetics.

6. Mitochondrial DNA Labeling

The mitochondrial genome is genetic material independent of the nuclear genome and is commonly found in eukaryotic cells. Mitochondrial DNA (mtDNA) is covalently closed, double-stranded, conserved, structurally simple, maternally inherited, and virtually free of recombination. mtDNA contains 13 protein-coding genes, 2 rRNA genes, 22 tRNA genes, and a noncoding control region. Typically, in fish mitochondrial DNA, the control region (D-loop) evolves at the fastest rate, and the rRNA genes evolve at the slowest rate. Since mitochondrial DNA follows matrilineal inheritance, its effective population size for detection is 1/4 of the nuclear DNA amphipathic mode of inheritance, and usually fewer samples of population genetic structure are detected using mitochondrial DNA-labeled genes for detection; thus, it is widely used in the identification of fish population structure.

For example, Liu et al. (2015a, b) used a 528 bp fragment of the mitochondrial COI gene sequence to analyze the genetic diversity of *Xenocypris argentea* populations in the Yangtze, Qiantang, and Heilongjiang rivers. The results showed that the haplotype diversity of the Yangtze River population ranged from 0.742 ± 0.053 to 0.991 ± 0.033, higher than that of the Heilongjiang River population at 0.731 ± 0.087 and the Qiantang River population at 0.552 ± 0.137. The interpopulation genetic differentiation index showed significant genetic differentiation among all three populations. Network trait analysis based on population haplotypes (Fig. 2.17) concluded that the Heilongjiang, Yangtze, and Qiantang river populations were three separate populations.

The maternally inherited properties of mitochondrial DNA reveal that mitochondrial DNA marker technology breaks through the limitations of Mendel's laws of inheritance to elucidate the nature of variation in terms of the primary structure of DNA. However, the genetic diversity information contained in a single mitochondrial marker has limitations, and the evolutionary rates of different mitochondrial genes vary among different organisms; thus, it is usually necessary to screen a larger number of marker genes and then use model selection to select more effective marker genes to determine

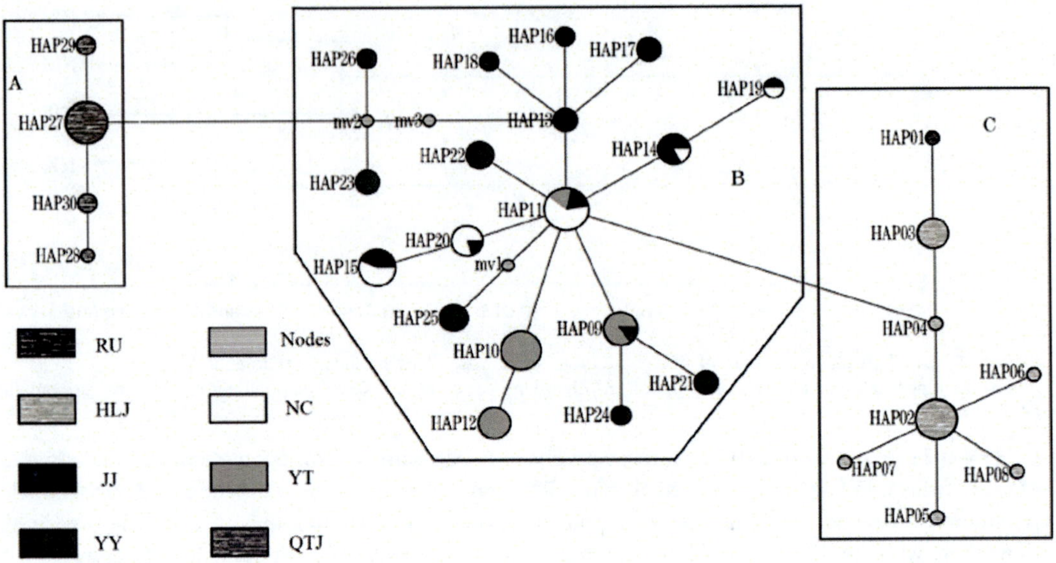

Fig. 2.17 Haplotype network relationship map of *Xenocypris argentea* populations (Liu et al. 2015a, b) *RU* Russian population, *HLJ* Heilongjiang population, *JJ* Jiujiang population, *YY* Yueyang population, *NC* Nanchang population, *YT* Yingtang population, *QTJ* Qiantang river population

the geographical distribution characteristics of populations.

2.3.5 Catch Statistics Method

The consistency, periodicity, and degree of variability in fishery conditions can also be used as a basis for stock identification by comparing catch statistics over long time series in various sea areas. Variation in a fish population is a property of that population and is an adaptation to the interaction between that fish population and its living environment. Because of the differences in the living environments of various fish populations, the pattern of population variation is not uniform. A catch is the result of fishing activities, and although the abundance or failure of a catch is influenced by both man-made and natural factors, changes in catches can indirectly reflect the trend in population changes and be used to identify stocks. For example, trout in the Kuril Islands are distributed on the eastern and western coasts of the Kamchatka Peninsula, and the abundance and failure of catches between the two coasts reverse yearly, so trout from the two coasts are considered to belong to different stocks.

2.3.6 Considerations for Stock Identification and Biological Sampling

The identification of populations is a complex and meticulous effort, especially sampling and identification, which should be representative and uniform. Moreover, a cautious approach should be taken in analyzing the information, focusing on the morphological method but relying on it mechanically, using as many identification methods as possible, comparing them with each other, and synthesizing them to avoid erroneous conclusions resulting from subjective and one-sided judgments; therefore, the following issues should be addressed in sampling and information analysis:

1. Because of the reproductive characteristics of fisheries resource stocks and the importance of

reproductive isolation as emphasized by the stock concept, sampling from spawning grounds is ideal, and collected samples must be grouped based on different spawning grounds according to the same criteria. In particular, composite samples should be taken from the beginning to the end of the reproductive period.
2. Sampling must take into account factors such as gear selectivity and catch variability.
3. In the determination of morphological characters, samples should be taken from fish with the same length range, with special attention to age and sex differences; variation due to living conditions should also be fully considered. Analysis of segmentation characteristics should be carried out with attention given to whether they are closely related to environmental conditions. Therefore, homogeneous information (e.g., same year and uniform fishing methods) must be considered for more reliable comparisons.
4. Ecological indicators must be analyzed given a full consideration of possible changes in their generation and living conditions.
5. A cautious approach to statistical analysis should be implemented, and a full account of biological significance should be considered in making judgments.
6. Determining stocks should be based on a comprehensive analysis of indicators to avoid the one-sidedness and chance associated with individual indicators.

2.4 Overview of Molecular Phylogeography

2.4.1 Establishment and Development of Molecular Phylogeography

With the continuous improvement in the theoretical basis of molecular phylogeography and the rapid development of molecular genetic markers, molecular phylogeography as a practice was born. Molecular phylogeography uses molecular biology techniques to reconstruct phylogenetic relationships within and across species, to explore the phylogeographic pattern of populations and closely related taxa, and to explain their evolutionary history. Molecular phylogeography is an approach that first obtains genetic information about the species or closely related groups of organisms under study through molecular genetic markers and then constructs phylogenetic trees among species, populations, and even individuals based on these molecular data. It then uses molecular clock theory and ancestral theory to infer the divergence time of ancestral populations and compares that time with the time of paleogeological and paleoclimatic events such as plate tectonic changes or ice age-interglacial changes to explore the formation mechanism of the phylogeographic pattern of the species in conjunction with their geographical distribution and to trace and reveal the evolutionary history of species.

Quaternary glacial-interglacial climate change had important effects on species formation, range evolution, and the formation of genetic patterns of extant species. During the glacial maxima, when northern Asia, Europe, and much of North America were covered by large ice sheets and sea level declines of up to 120–140 m occurred, many organisms underwent large changes in range areas and population sizes, eventually migrating to several compressed ice age refuges, and postglacial population expansion occurred. This scenario resulted in the present geographic distribution pattern. These spatial and temporal distributions of species leave genetic imprints that can be obtained through molecular genetic markers, which in turn can trace and reveal the evolutionary history of species. Since the establishment of molecular phylogeography, studies of the phylogeographic patterns of terrestrial plants and animals have been concentrated in North American and European regions. The field is also widely used in marine organisms, mainly marine fishes, such as the testing of hypotheses of marine biogeography and the detection of population history dynamics and their influencing factors.

In recent years, research on molecular phylogeography has rapidly progressed and become a popular area of research internationally. For example, the uplift of the Qinghai-Tibet Plateau had an important impact on the shaping of biodiversity and speciation history in the region, providing preconditions for the application of the isolation hypothesis, while global large-scale climate change has had an important impact on the habitat environments of organisms, causing them to spread to suitable habitats and form their current spatial and temporal distribution patterns. Molecular phylogeography is also gradually expanding from single species to regional proximate taxa, which facilitates speculation on the systematic evolution of proximate taxa.

Molecular phylogeography has been widely used because of its ability to reconstruct phylogenetic relationships at the intraspecific and supraspecific levels and thus to interpret the evolutionary history of species, but it also has shortcomings. Its shortcomings are mainly reflected in the following two aspects: (1) the theoretical basis of molecular phylogeography is not well developed, and (2) the defects in molecular genetic markers are problematic.

2.4.2 Case Study: Molecular Phylogeography of Jumbo Flying Squid in the Eastern Pacific

(1) Materials and Methods

1. Experimental Materials and Extraction of Genomic DNA

Jumbo flying squid were harvested from equatorial waters and off Peru, stored in cabin coolers, and transported back to the laboratory (Table 2.1). Mantle muscle tissue was taken and placed in 95% ethanol and stored at −20 °C to serve as a backup sample. Genomic DNA extraction was performed using a tissue/cellular genomic DNA rapid extraction kit.

2. PCR Amplification

The primers for COI gene amplification were self-designed: COIF, 5′-ATCCCATGCAGGCCCTTCAG-3′, and COIR, 5′-GCCTAATGCTCAGAGTATTGGGG-3′. *Cytb* gene amplification primers were cited from Yan et al. (2011): CytbF, 5′-ACGCAAAATGGCATAAGCGA-3′, and CytbR, 5′-AGTTGTTCAGGTTGCTAGGGGA-3′. The PCR amplification reaction systems were all 25 μL, including 10× PCR Buffer at 2.5 μL, *Taq* DNA polymerase (5 U/μL) at 0.2 μL, dNTP (2.5 mmol/L each) at 2 μL, upstream and downstream primers (10 μmol/L) at 0.6 μL each, DNA template at 20 ng, and ddH_2O to make up the volume. The PCR procedures were the following: predenaturation at 94 °C for 2 min, denaturation at 94 °C for 30 s, annealing at 58 °C for 45 s, extension at 72 °C for 45 s and 35 cycles, and final extension at 72 °C for 2 min.

3. Purification and Sequencing of PCR Products

PCR products were separated by 1.2% agarose gel electrophoresis, purified with a Biospin Gel Extraction Kit, and sequenced in both directions on an ABI3730 Genetic Analyzer.

4. Data Analysis

Sequencing results were compared using ClustalX 1.83 software and manually proofread. The base composition of the DNA sequences was determined using Statistics in MEGA 4.0 software, and net genetic distances were calculated using the TrN + G model. Genetic diversity

Table 2.1 Information on sample collection of jumbo flying squid (Liu 2014)

Sea area	Sampling locations	Sampling time	Average mantle length/cm	Average body weight/g	Number of samples
Equatorial sea	3°N–5°S, 114° ~ 120°W	2011-12 to 2012-04	33.28 ± 9.65	1378.26 ± 1096.57	33
Off Peru	10°–11°S, 82° ~ 84°W	2011-09 to 2011-10	24.65 ± 1.58	436.41 ± 136.98	33

parameters such as haplotype number, haplotype diversity index, nucleotide diversity index (π), and mean nucleotide difference number (k) were calculated by DnaSP 4.10 software. The relationship between haplotypes and the number of nucleotide differences was calculated by Arlequin 3.01 software by constructing a minimum spanning tree to reflect the linkage between different haplotypes, and the genetic differentiation coefficient F_{st} between populations and its significance were calculated using this software (replicates 1000). Tajima's *D* and Fu's Fs neutrality tests and nucleotide mismatch distribution were used to detect the historical dynamics of the squid population. The historical population expansion time was estimated using the parameter τ, which was transformed into the actual expansion time by the equation $\tau = 2ut$, where u is the mutation rate for the entire sequence length under study and t is the time since the beginning of the population expansion to the present. Nucleotide divergence rates for *Cytb* genes ranged from 2.15% to 2.6% per million years.

(2) Results

1. Sequence Analysis

The amplification product of the *Cytb* gene fragment was obtained by PCR amplification, purified, then sequenced, and sequence matched to obtain a comparable sequence of 724 bp. The average contents of the A, T, G, and C bases were 43.97%, 23.61%, 12.25%, and 20.17%, respectively, in all analyzed sequences, with the A + T content (67.58%) being significantly higher than the G + C content (32.42%) (Table 2.2). Eighteen variant sites were detected in the *Cytb* gene fragment, including nine single-base variant sites and nine parsimony informative sites. Conversions and reversals were 16 and 2, respectively, with no insertions or deletions. These variant loci defined a total of 21 haplotypes, of which haplotypes H3, H4, and H7 were shared by all populations, and the remaining haplotypes were owned by single individuals (Table 2.3).

Comparable sequences of 622 bp of the COI gene fragment were obtained according to the *Cytb* gene sequence analysis method. The average contents of the A, T, G, and C bases were 27.69%, 36.67%, 15.39%, and 20.25%, respectively, in all analyzed sequences, with the A + T content (64.36%) being significantly higher than the G + C content (35.64%) (Table 2.2). Twenty variant sites were detected in the COI gene fragment, including 14 single-base variant sites and 6 parsimony informative sites. The conversions and reversals were 17 and 3, respectively, with no insertions or deletions. These variant loci defined a total of 19 haplotypes, with haplotypes H1, H5, and H6 being haplotypes shared by all populations (Table 2.4). Haplotype sequence divergence values were low, with nine of the remaining 18 haplotypes differing by only one nucleotide when compared to haplotype H1.

(2) Population Genetic Diversity

The total number of haplotypes, haplotype diversity index, nucleotide diversity index, and mean number of nucleotide differences for the two populations obtained based on all sequences of the *Cytb* gene fragment were 21, 0.767 ± 0.047, 0.00224 ± 0.00144, and 1.548, respectively. The results obtained based on all sequences of the COI gene fragment were 19, 0.743 ± 0.055, 0.00267 ± 0.00178, and 1.658, respectively. As seen from Table 2.5, the haplotype diversity index, nucleotide diversity index, and mean number of nucleotide differences were higher in the equatorial marine population than in the Peruvian offshore population.

Table 2.2 Sequence composition of COI and *Cytb* gene fragments in jumbo flying squid (Liu 2014)

			Base content (%)					
Gene (barcode)	Clip length (bp)	Number of gene sequences	A	T	G	C	A + T	G + C
COI	622	67	27.69	36.67	15.39	20.25	64.36	35.64
Cytb	724	67	43.97	23.61	12.25	20.17	67.58	32.42

Table 2.3 *Cytb* haplotypes of jumbo flying squid and their distribution in the population (Liu 2014)

	Variant site																		Distribution of haplotypes		
	0	0	1	1	1	1	1	2	2	3	3	4	4	4	4	5	6	6			
	8	9	0	1	5	8	9	5	9	3	4	1	7	8	8	8	1	1			
Haploid (type)	8	9	0	3	1	7	8	0	2	7	3	8	5	7	9	0	3	9	Equatorial sea	Off Peru	*n*
H1	G	A	A	T	A	G	T	G	C	T	A	A	T	T	A	G	C	C	1		1
H2	–	G	–	–	G	–	–	–	–	–	–	G	–	–	–	–	–	–	1		1
H3	–	–	–	A	G	–	–	–	–	–	–	G	–	–	–	–	–	–	5	5	10
H4	–	–	–	A	G	A	–	–	–	–	–	–	–	–	–	–	–	–	14	16	30
H5	–	–	–	A	G	A	–	–	–	–	–	–	–	–	G	–	–	–	1		1
H6	–	–	–	A	G	–	–	–	–	–	–	G	–	–	–	A	–	–	1		1
H7	–	–	–	A	G	A	–	–	–	–	–	G	–	–	–	–	–	–	7	2	9
H8	–	–	–	A	G	–	–	–	–	–	–	–	–	–	–	–	–	–	1		1
H9	–	–	–	A	G	A	–	–	T	C	–	G	–	–	–	–	–	–	1		1
H10	–	–	–	A	G	A	A	–	–	–	–	–	–	C	–	–	–	–	1		1
H11	–	–	–	A	G	A	A	–	–	–	–	–	–	–	–	–	–	–	1		1
H12	–	–	–	A	G	A	–	–	–	–	G		–	–	–	–	–	–		1	1
H13	A	–	–	A	G	A	–	–	–	–	–	G	–	–	–	–	–	–		1	1
H14	–	–	–	A	G	A	–	–	–	–	G	G	–	–	–	–	–	–		1	1
H15	–	–	–	A	G	–	–	–	–	–	–	G	C	–	–	–	–	–		1	1
H16	–	–	G	A	G	A	–	–	–	–	–	G	–	–	–	–	T	T		1	1
H17	A	–	–	A	G	A	–	–	–	–	–	–	–	–	–	–	–	–		1	1
H18	–	–	–	A	–	–	–	–	–	–	–	G	–	–	–	–	–	–		1	1
H19	–	–	–	A	G	A	–	–	–	C	–	G	–	–	–	–	–	–		1	1
H20	–	–	–	A	G	A	–	–	–	–	–	–	–	C	–	–	–	–		1	1
H21	–	–	–	A	G	A	–	A	–	–	–	G	–	–	–	–	–	–		1	1

(3) Genetic Differentiation Between Populations

The minimum spanning tree of *Cytb* and COI haplotypes shows that there were no significantly divergent haplotype taxa within the jumbo flying squid population, the haplotypes were linked by single- and multistep mutations, and the minimum spanning tree had a starlike structure, suggesting that stem catfish may have experienced population expansion events (Fig. 2.18). Combining the *Cytb* gene sequencing results of the Costa Rican population of stem catfish to construct haplotype neighbor-joining evolutionary trees for the three geographic populations (Fig. 2.19), there were no significantly divergent haplotype taxa between the Peruvian offshore population and the equatorial marine population, and the Costa Rican offshore population had two haplotypes forming one significantly divergent haplotype taxon. Analysis of the genetic differentiation coefficient F_{st} between the two populations showed that there was no significant genetic differentiation between the equatorial marine population and the Peruvian offshore population (*Cytb*: $F_{st} = 0.013\ 76$, $P > 0.05$; COI: $F_{st} = 0.021\ 60$, $P > 0.05$) (Table 2.6).

(4) Group History Dynamics

Tajima's *D* and Fu's *Fs* neutral tests based on the *Cytb* gene fragment sequences had negative *D* and Fs values, and both statistical tests were significant (Table 2.7). The results of the nucleotide unpaired distribution analysis showed that the *Cytb* haplotype nucleotide unpaired distribution had a single-peaked type (Fig. 2.20), and the observed values did not significantly deviate from the simulated values (Hri = 0.0317; $P = 0.673$). The above results suggest that jumbo flying squid may have experienced a recent population expansion event.

Table 2.4 Haplotypes of jumbo flying squid COI and their distribution in the population (Liu 2014)

	Variant site																				Distribution of haplotypes		
	0	0	0	1	1	1	2	2	2	2	3	3	4	4	4	5	5	5	5	5			
	4	6	8	4	6	7	6	6	7	7	4	8	0	7	9	1	2	2	3	3			
Haploid (type)	7	4	2	2	0	9	5	8	1	2	1	9	0	8	1	5	3	9	5	8	Equatorial sea	Off Peru	*n*
H1	C	T	C	G	T	C	G	A	A	G	G	G	A	A	G	G	G	T	C	A	14	18	32
H2	–	–	–	–	–	–	–	G	–	–	–	–	–	–	–	–	–	–	–	–		1	1
H3	–	–	–	–	–	T	–	–	–	–	–	–	–	–	–	–	–	–	–	–		1	1
H4	–	–	–	–	–	–	–	–	–	–	–	–	–	–	–	–	–	–	–	G		1	1
H5	–	–	–	–	–	–	–	–	–	–	–	–	–	–	–	–	A	–	–	–	4	4	8
H6	–	–	T	–	–	–	–	–	–	–	–	–	–	–	–	–	–	–	–	–	2	2	4
H7	–	–	–	–	–	–	–	G	–	–	–	A	–	–	A	–	A	–	–	–		2	2
H8	–	–	–	–	–	–	–	G	G	–	–	A	–	–	–	–	–	–	–	–		1	1
H9	–	–	–	–	–	–	–	–	–	–	–	–	–	–	–	–	–	–	T	–		1	1
H10	–	–	–	A	–	–	–	–	–	–	–	–	–	–	–	–	–	–	–	–		1	1
H11	–	–	–	–	–	–	–	–	–	–	–	–	–	–	–	–	–	C	–	–		1	1
H12	T	–	–	–	–	–	–	–	–	–	–	–	–	–	–	–	A	–	–	–	2		2
H13	–	–	–	–	–	–	–	–	–	–	–	–	G	–	–	–	A	–	–	–	2		2
H14	–	–	–	–	–	–	–	G	–	A	–	–	–	–	–	–	A	–	–	–	2		2
H15	–	–	–	–	–	–	–	G	–	A	C	A	–	–	A	C	–	–	–	–	2		2
H16	–	–	–	–	–	–	–	–	–	–	–	–	G	–	–	–	–	–	–	–	2		2
H17	–	A	–	–	C	–	–	–	–	–	–	–	–	–	–	–	–	–	–	–	2		2
H18	–	–	–	–	–	–	A	G	–	–	–	–	–	–	–	–	–	–	–	–	1		1
H19	–	–	–	–	–	–	–	–	–	–	–	–	–	G	–	–	A	–	–	–	1		1

Table 2.5 Genetic diversity of two geographic populations of jumbo flying squid based on *Cytb* and COI gene fragment sequences (Liu 2014)

Gene (barcode)	Community	Number of samples	Haplotype number (math.)	Haplotype diversity index	Nucleotide diversity index (π)	Average number of nucleotide differences (k)
Cytb	Equatorial sea	34	11	0.783 ± 0.055	0.002 39 ± 0.001 36	1.568
	Off Peru	33	13	0.752 ± 0.075	0.002 14 ± 0.001 15	1.549
	Total the total	67	21	0.767 ± 0.047	0.002 24 ± 0.001 44	1.548
COI	Equatorial sea	34	11	0.815 ± 0.060	0.003 57 ± 0.002 25	2.223
	Off Peru	33	11	0.663 ± 0.090	0.001 76 ± 0.001 12	1.095
	Total	67	19	0.743 ± 0.055	0.002 67 ± 0.001 78	1.658

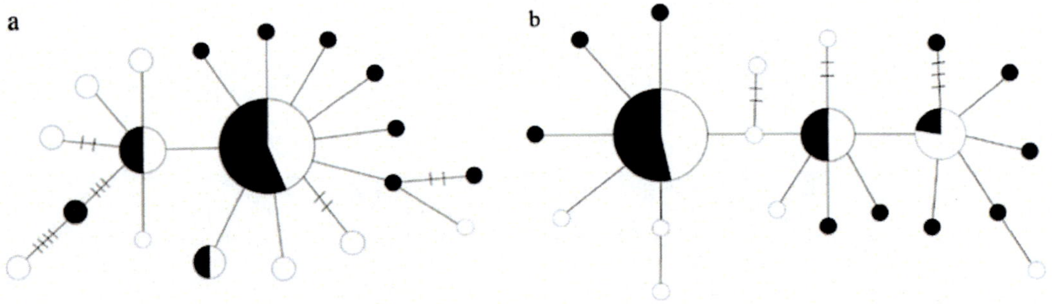

Fig. 2.18 Minimum spanning tree of haplotypes of jumbo flying squid COI (**a**) and *Cytb* (**b**) (Liu 2014)
Note: The area of the circle is proportional to the haplotype frequency, and the short underline represents the number of nucleotide substitutions between haplotypes. Population: ■ off Peru; □ equatorial seas

Based on a τ value of 3.588 for the distribution of *Cytb* haplotype nucleotide unpairs in jumbo flying squid, the expansion event of the jumbo flying squid population was calculated to have occurred between 125,000 and 151,000 years ago, during the late Pleistocene, when global climatic and marine environmental changes had a significant impact on the spatial distribution patterns of many marine organisms. The presence of one significantly divergent haplotype taxon branch within jumbo flying squid population suggests that the Costa Rican population of jumbo flying squid was segregated from the Peruvian offshore and equatorial marine populations during the Pleistocene and that the population genetic structure patterns of the Peruvian offshore and equatorial marine populations were primarily driven by current factors and less influenced by historical factors.

2.5 Population Growth and Regulation Process

Internal and external factors affect the growth of fish populations. Internal factors refer mainly to fecundity and mortality. External factors include biotic and abiotic factors, with biotic factors referring mainly to competitors or predators and abiotic factors referring mainly to physical

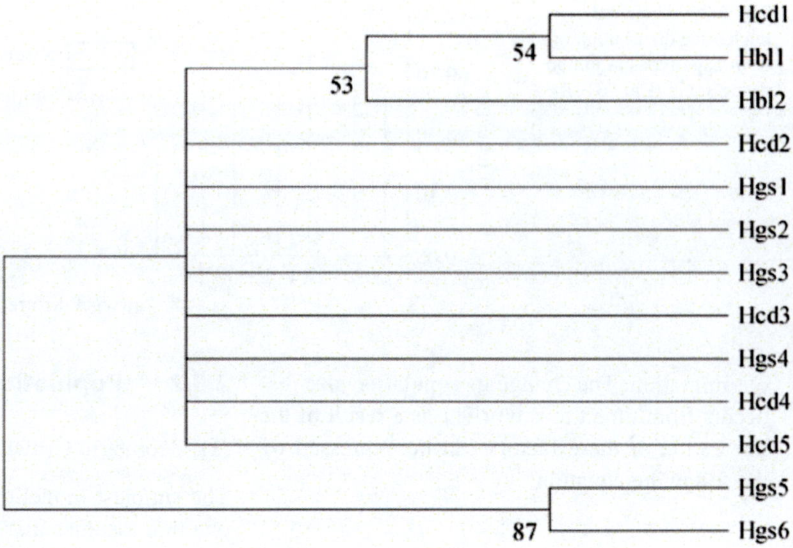

Fig. 2.19 Neighbor-joining evolutionary tree constructed based on the *Cytb* haplotype of jumbo flying squid (Liu 2014) Note: Hcd, Hbl, and Hgs are the haplotypes of the equatorial sea population, the Peruvian offshore population, and the Costa Rican offshore population, respectively

Table 2.6 Genetic differentiation coefficients F_{st} between two geographical groups of jumbo flying squid (Liu 2014)

Community	Equatorial sea	Off Peru
Equatorial sea	–	0.021 60 ($p > 0.05$)
Off Peru	0.013 76 ($p > 0.05$)	–

Note: The diagonal line shows the result of *Cytb* gene fragment sequence analysis, and the diagonal line shows the result of COI gene fragment sequence analysis

Table 2.7 Neutral test results of the *Cytb* gene in jumbo flying squid (Liu 2014)

		Tajima's *D*		Fu's *Fs*	
Gene (loanword)	Community	*D*	*P*	*Fs*	*P*
Cytb	Equatorial sea	−1.330	0.091	−27.656	0.000
	Off Peru	−1.502	0.061	−27.590	0.000
	Total	−1.774	0.027	−17.926	0.000

Note: Bars are observations, and curves are the expected distributions under the population expansion model

environmental constraints, such as light, water temperature, and currents. Theoretically, it is possible to estimate or predict the growth rate of fish populations if the above internal and external factors and their influencing effects are known. There are many models used to describe the growth of fish populations, mainly including geometric growth models, exponential growth models, and logistic growth models.

2.5.1 Fish Population Growth Processes

To explain the process of population size growth, we assume that external factors have no effect on population size growth, explore only internal factors affecting population size growth, and assume population size increases due to birth and in-migration and decreases due to death and

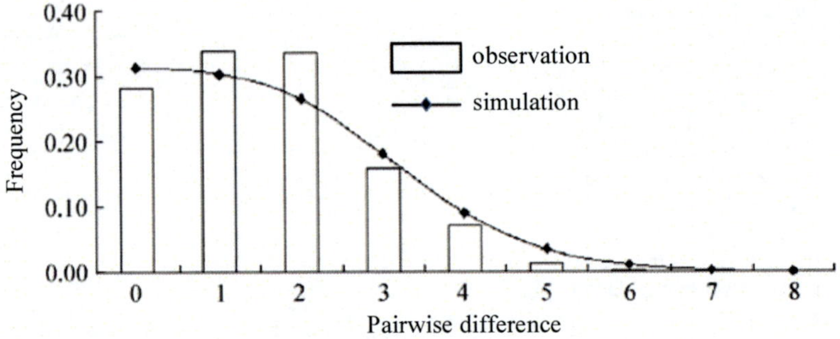

Fig. 2.20 Unpaired nucleotide distribution of *Cytb* haplotypes in jumbo flying squid (Liu 2014)

out-migration. The change in population size that occurs from time t to time $t + 1$ as a result of the interaction of these factors can be expressed by the following equation:

$$\Delta N = N_{t+1} - N_t = B + I - D - U \quad (2.5)$$

where B, I, D, and U are the number of populations being born, moving in, dying, and moving out at time t to $t + 1$, respectively; N_t and N_{t+1} are the number of populations at time t and $t + 1$, respectively. For a single population with essentially no in-migration and out-migration, I and U may take on the value of zero, and Eq. (2.5) may become the following:

$$N_{t+1} - N_t = B - D \quad (2.6)$$

The total number of births and deaths in the population are both a function of the number of individuals in the population. Thus, we have $B = b_{Nt}$ and $D = d_{Nt}$, where b and d are the reproductive rate and mortality rate of the population, respectively. That is, b is the number of new individuals who can be produced or reproduced per individual, and d is the probability that an individual will die in a given time. Then, Eq. (2.6) can be changed to the following:

$$N_{t+1} - N_t = (b - d)N_t \quad (2.7)$$

From Eq. (2.7), the population will increase if the reproductive rate is greater than the mortality rate; if the mortality rate is greater than the reproductive rate, then the population will decrease. However, a population does not grow indefinitely and will tend to a finite value due to limitations such as prey and space.

2.5.2 Population Growth Model

(1) Geometric Growth Model

The simplest model of population size growth is one that assumes that the rate of change in population size is constant and independent of the density of the distribution; this model of population size growth is called the geometric growth model. In such a population, reproduction occurs only once per generation, and the mother dies after reproduction. Assuming that each average individual produces R_0 offspring, R_0 is defined as the net growth rate per generation so that R_0 is the following:

$$R_0 = N_{t+1}/N_t \quad (2.8)$$

Putting Eq. (2.8) in order, the population size in generation 1 is $N_1 = R_0 \times N_0$ and that in generation 2 is $N_2 = R_0 \times N_1 = R_0^2 \times N_0$ so that the population size in generation t is the following:

$$N_t = R_0^t \times N_0 \quad (2.9)$$

If R_0 is greater than 1, then the population size will increase over time; if R_0 is less than 1, then the population size will decrease over time; when R_0 is equal to 1, the population size will remain constant.

(2) Exponential Growth Model

In some populations, individuals reproduce almost continuously, with no special reproductive periods. In this case, the change in the size of the population can be expressed by the following differential equation:

$$dN/dt = (b - d) \times N \tag{2.10}$$

where dN/dt represents the change in population size over a very short time interval. The difference between the reproduction rate and the mortality rate is denoted by r; i.e., $r = b - d$. Then, r can be called the endogenous growth rate of the population. Then, Eq. (2.10) can be expressed as the following:

$$dN/dt = r \times N \tag{2.11}$$

From Eq. (2.11), it can be seen that if r is greater than 0, then the population size will increase; if r is less than 0, then the population size will decrease; if $r = 0$, then the population size will remain the same.

In addition, Eq. (2.11) can be expressed as follows:

$$N_t = N_0 \times e^{rt} \tag{2.12}$$

where N_0 is the population size at moment 0.

(3) Logistic Model

In geometric and exponential growth models, population growth will continue to infinity when R_0 is greater than 1 or r is greater than 0, a phenomenon that is not common in nature. In general, population size will tend to a finite value, limited by food, space, or other resources in the environment, a limit also known as the load capacity or carrying capacity. Certain environmental conditions will support a certain number of individuals. Additionally, the size of the carrying capacity of a natural population is determined to a large extent by the level of resources available under certain environmental conditions. The population growth of populations such as fish essentially falls under the logistic model.

Load capacity can be applied in a population growth model where the population size increases while the growth rate decreases. When the population size is equal to the load capacity, population growth stops, and the population size remains constant. If the load capacity is denoted by K_1, then the exponential growth model can be changed to a logistic model (e.g., Fig. 2.21):

$$\frac{dN}{dt} = rN(1 - \frac{N}{K_1}) = F(N) \tag{2.13}$$

where N is the population size, r is the population endogenous growth rate, and K_1 is the load capacity.

Equation (2.13) is the logistic growth model and forms the S-shaped curve in Fig. 2.20. By solving this problem, the logistic growth model can also be expressed as the following:

$$N_t = K_1/[1 + (K_1/N_0 - 1)e^{-rt}] \tag{2.14}$$

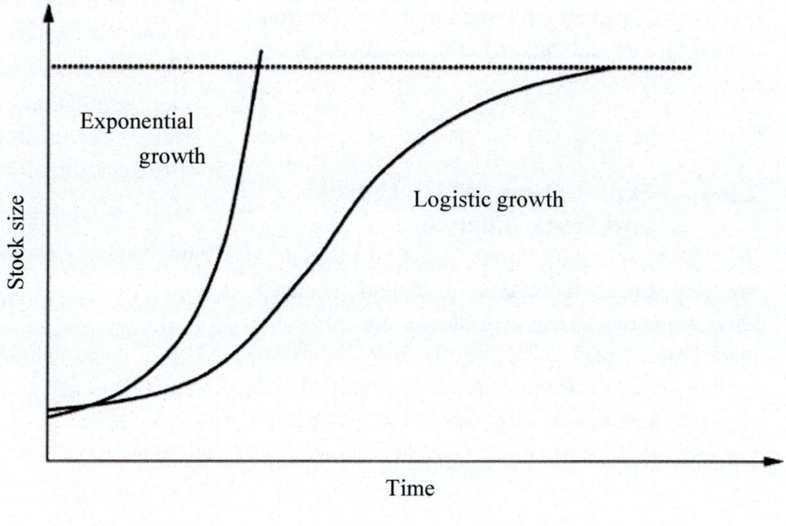

Fig. 2.21 Exponential and logistic growth curves

2.5.3 Basic Characteristics of the Dynamics of Natural Fish Population Sizes Under Limited Environmental Conditions

The evolutionary process of any fish resource development is limited by anthropogenic regulation, its own renewal function, and environmental capacity, with the first limitation being socioeconomic factors and the third limitation being natural factors. In addition to the initial size, growth rate, mortality, and time mentioned earlier, additional natural factors are the following: (1) interspecific competition, where various organisms compete with each other for other biological resources needed to sustain their growth, which is known as "survival of the fittest." The result of this competition is the rapid depletion of some species and the development of others. (2) Spatial limitation is another natural factor. In biology, an increase in population density will lead to a reduction in individual space, thus changing a species proliferation rate; when the population density reaches a certain level, proliferation stops. (3) Survival conditions are the third factor. Even if space is infinite, the quantitative growth of some resources can change the exponential growth trend they initially exhibit due to the nonfulfillment of certain necessary natural factors (e.g., climate). Thus, even if the effects of anthropogenic regulatory activities are excluded, the change in the quantity of fish resources over time is governed by a limited number of environmental factors and thus exhibits a limited growth trend.

2.5.4 Exploitation of Fish Stocks and Stock Balance

For fish stocks, economic exploitation means harvesting, but changes in stocks are not equal to harvesting rates and are regulated by biological growth or renewal capacity. It is assumed that changes in fish population size when the stock is not affected by anthropogenic factors are characterized by the following equation:

$$\frac{dN}{dt} = F(N) = rN\left(1 - \frac{N}{K_1}\right) \qquad (2.15)$$

The derivation of Eq. (2.15) yields the population level at which the growth rate of fish stocks is maximized. That is, $N = K_1/2$ when its fish stock growth rate reaches a maximum of $rK_1/4$.

When there is exploitation of fish stocks, the change in population size is influenced by the rate of fishing (h), and the equation is the following:

$$\frac{dN}{dt} = F(N) - h(t) \qquad (2.16)$$

The relationship between fishing behavior and population growth rates is analyzed in purely biological terms.

1. When the capture rate is h_1, $h_1 < \frac{1}{4}rK_1$, Eq. (2.16) has two equilibria N_1 and N_2 (Fig. 2.21), and the population dynamics depend on the size of its stock N_t. If N_t lies between N_1 and N_2, then $\frac{dN}{dt} > 0$, N_t will tend to N_2, which also occurs when $N_t = K_1$; if N_t lies to the left of N_1, then $\frac{dN}{dt} < 0$, N_t will tend to 0, or the predetermined time N_t will be 0.
2. When the catch rate is h_{max}, $h_{max} = rK_1/4$; then, Eq. (2.16) has a unique equilibrium point at $N^* = K_1/2$ (Fig. 2.21), at which $\frac{dN}{dt} = 0$, and the catch rate at that point is called the maximum sustainable yield (MSY).
3. When the capture rate is h_2, $h_2 > \frac{1}{4}rK_1$, then $\frac{dN}{dt} < 0$. For any stock level N_t of the resource, the population size will tend to zero (Fig. 2.22), a situation commonly referred to as overexploitation.

Now, assuming that the catch is proportional to the level of fish stocks, we have

$$h = qEN \qquad (2.17)$$

where q is the fishing coefficient, E is the fishing intensity, and N is the population size.

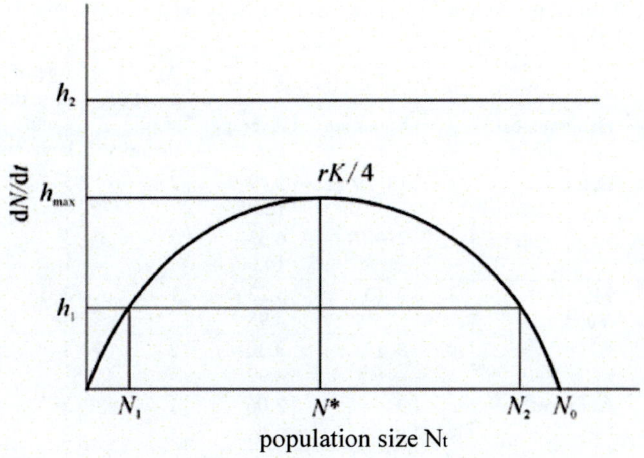

Fig. 2.22 Dynamic effects of fishing on fish population size

Substituting Eq. (2.17) into Eq. (2.16) yields the following:

$$\frac{dN}{dt} = F(N) - h(t) = rN\left(1 - \frac{N}{K_1}\right) - qEN$$

Let $\frac{dN}{dt} = 0$, then $N^* = K_1(1 - Eq/r)$.

The equilibrium catch or sustained yield $Y = h$ corresponding to the fishing intensity E can be calculated by the following equation:

$$Y = qEK_1\left(1 - \frac{Eq}{r}\right) = aE - bE^2 \qquad (2.18)$$

of which $a = qK_1, b = K_1q^2/r$.

Equation (2.18) was proposed by biologist M.B. Schaefer and is therefore also known as the Schaefer model. Solving for Eq. (2.18) yields a MSY of $\frac{a^2}{4b}$, which corresponds to a fishing intensity of $E_{MSY} = \frac{a}{2b}$.

2.6 Case Studies of Stock Identification

2.6.1 Geographical Stock Division of Large Yellow Croaker Offshore Chinese

The formation and differentiation of geographical fish populations are closely related to regional environmental conditions, and there are obvious differences in the morphological and ecological characteristics and other characteristics of different populations. The main biological characteristics of the three geographic populations of large yellow croaker offshore Chinese are summarized in Table 2.8. The relationship between their geographical populations and ecological environment is analyzed as follows.

1. Population structure of the Daiqu population in relation to geographic latitude. The Daiqu population, distributed in northern Zhejiang province, has the longest life span, a late sexual maturity age (especially at the age of substantial sexual maturity), a complex composition, and a stable presence. The Naozhou population, living west of the Pearl River Estuary in the South China Sea, has a short life span, the earliest sexual maturity, a simple composition, and a larger proportion of the recruitment group. In contrast, the characteristics of the Min-Yuedong population, which is distributed in the sea area from northern Fujian to the Pearl River Estuary, are moderate in comparison to those of the other two populations. Thus, the variation in longevity and composition of the abovementioned geographic populations also shows a general pattern of temperature-dependent changes affecting animals; i.e., the geographic populations of large yellow croaker inhabiting distinctly different latitudinal ranges along the

Table 2.8 Main biological indicators of the geographic populations of large yellow croaker (Chen 2014; Chen and Liu 2017)

Geographical populations	Sex	Maximum age	Average age	Age of onset of sexual maturity	Age of onset of sexual maturity in large numbers	Number of reproductive fish age groups	Remaining groups (%)	Catch (relative weight)
Daiqu	♀	18–29	5.49–12.98	2	3–4	17–24	80–85	100
	♂	21–27	6.33–14.00	2	3	20–24		
Min-Yuedong	♀	9–13	3.23–4.98	1–1	2–3	8–12	60–65	25
	♂	8–17	3.30–4.92	2	2	8–16		
Naozhou	♀	9	3.06	1	2	7	60–65	1
	♂	8	2.94	1	2	8		

Chinese coast have a longer population life span, delayed sexual maturity of generations, and more complex population composition as the latitude of its range increases, and the stability of population variation increases.

2. Relationship between geographic populations of large yellow croaker and its environment. The numbers of the three geographic populations of large yellow croaker are linked to the size of their spawning grounds and, to a lesser extent, to the size of their living areas and the amount of major river runoff. That is, the Daiqu population is the most extensive in size, the Naozhou population is the narrowest, and the Min-Yuedong population size is in between those of the other two. Third, in terms of the species compositions of the fish fauna of the East China Sea and the South China Sea, the South China Sea species composition is more than twice as great as that of the East China Sea. Fourth, in terms of proliferative capacity and stability of population changes, the Daiqu population is characterized by both a complex and more stable population structure with increasing longevity and a relatively rapid rate of replenishment. In summary, the disparate differences in species numbers, the length of the species life span, and degree of compositional complexity or simplicity among the three geographic populations that comprise large yellow croaker are a manifestation of the adaptation of each population to the oceanographic conditions and geographic characteristics that have occurred over the course of history in the sea where each population lives.
3. Geographic variation in morphological characteristics of large yellow croaker. Most of the segmentation characteristics of large yellow croaker, particularly the number of gill rakers and swim bladder lateral branches, and the mean number of body measurements, such as eye diameter, caudal peduncle height, and body height, exhibit significant geographic variation from north to south in a latitude-parallel order (Table 2.9). The number of dorsal fin spines, pyloric blind sacs, and lateral swim bladder branches gradually increases from north to south, while the number of gill rakers, vertebrae, and anal fin rays decreases in the opposite order from north to south; the size of eye diameter is greater in the first northern taxon than in the southern taxon; and the two characters, caudal peduncle height and body height, gradually increase from north to south.

Based on information from the three major reproductive populations of large yellow croaker, there is a relationship between the variation in mean eye diameter size of the various populations and the transparency of the seawater in the area of distribution; i.e., populations with larger mean

Table 2.9 Differences in morphological characteristics of the three populations of large yellow croaker (Chen 2014; Chen and Liu 2017)

Features	Number of dorsal fin spines	Number of blind pyloric sacs	Number of lateral branches of swim bladder		Number of gill rakes	Vertebrae count	Number of anal fins
			Left side	Right side			
Daiqu	9.91	15.12	29.81	29.65	28.52	26.00	8.07
Min-Yuedong	9.96	15.20	30.57	30.46	28.02	25.99	8.04
Naozhou	9.96	15.29	31.74	31.42	27.39	25.98	8.01
Features	**Number of pectoral fins**	**Number of dorsal fins**	**Eye diameter/head length**	**Tailstock height/tailstock length (%)**	**Body height/L (%)**	**D-A/L (%)**	
Daiqu	16.82	32.53	20.20	27.80	25.29	46.31	
Min-Yuedong	16.78	32.64	19.19	28.42	25.58	46.46	
Naozhou	16.68	32.27	19.40	28.97	25.96	47.02	

Table 2.10 Relationship between eye size and seawater transparency for three populations of large yellow croaker (Chen 2014; Chen and Liu 2017)

Populations (of the species)	Eye diameter/head length %		Water transparency at major spawning grounds	
	Fluctuation range	average	Fluctuation range (m)	Key values (m)
Daiqu	17–23	20.16	<0.5–1.5	<0.5
Min-Yuedong	17–21	19.19	0.5–1.1	0.6–0.8
Naozhou	17–22	19.40	0.5–2.0	1.0–2.0

eye diameters are located in areas with lower seawater transparency (Table 2.10).

The three geographic populations of large yellow croaker differ somewhat in their statistical patterns, and accordingly, they are also ecologically distinct. The marine environmental conditions in the main habitat areas of the three geographic populations account for the differences among the populations. The first environmental condition that varies is the climate, with climatological differences arising from the different latitudes where the populations are located, i.e., the difference in marine climate from temperate-subtropical in the northernmost range to subtropical-tropical in nature in the southernmost range. The second environmental condition that differs results from the influence of nearshore marine currents and river runoff in China. The main distribution area of the first population is generally from the southern or central Yellow Sea in the north to north of the Taiwan Strait in the south, and most of its area is influenced by Yangtze River runoff. The distribution area of the second population is approximately south of Yushan Island to the mouth of the Pearl River, and all this area is directly or indirectly influenced by the marine conditions of the Taiwan Strait. The distribution area of the third population is the area along the South China Sea from the mouth of the Pearl River west to the Qiongzhou Strait, which is characterized by the nature of the inlet.

4. In summary, the main morphological and ecological characteristics of the three populations of large yellow croaker differ markedly, showing a continuous gradient of geographic variation, with more pronounced differences between the more geographically isolated Daiqu and Naozhou populations and the transitional nature of most of the characteristics of the intermediate Min-Yuedong population. The main morphological features of the Daiqu population are the higher number of gill rakers, lower number of bladder lateral branches, larger eye diameter, and relatively lower caudal stalk and body than those of the two southern populations. In terms of ecological characteristics, those in the Daiqu population have the longest life span, later sexual maturity, the most complex composition, a relatively stable population size, and a predominantly spring reproductive period. The main morphological characteristics of the Naozhou population are a lower number of gill rakers, a higher number of swim bladder lateral branches, a smaller eye diameter, and a relatively higher caudal peduncle and body. Ecologically, this population has the shortest life span, early sexual maturity, simpler composition, and a predominantly autumn reproductive period. On the other hand, the Min-Yuedong population has morphological and ecological characteristics that are intermediate between those of the Daiqu and Naozhou populations (only the eye diameter is slightly smaller than that of the Naozhou population), and in terms of reproductive period, it occurs in spring for the northern Min-Yuedong population and in autumn for the southern Min-Yuedong population.

2.6.2 Stock Identification of Jumbo Flying Squid in the Southeastern Pacific Based on Landmark Point Analysis

(1) Materials and Methods

1. Acquisition of Two-Dimensional Images

The collected samples of the jumbo flying squid were sent to the laboratory for thawing, and the left and right otoliths were removed with forceps and placed into 1.5 ml centrifuge tubes and numbered for storage. Organic material and debris were removed from the surface of the otoliths using an ultrasonic cleaner, and one side of the otolith was uniformly selected with the convex side facing upward and placed under an Olympus light microscope (objective $\times$ 4, eyepiece $\times$ 10) $\times$ 40$\times$ to take pictures with a CCD. After image acquisition, the otoliths were processed using Photoshop CS5 to ensure the clarity of the external morphology of the otoliths.

2. Selection of Landmark Points

The landmark points were biologically divided into three major categories, and a total of 14 landmark points were selected by combining previous studies on otolith morphology: 2, 8, 10, 11, and 13 are defined as type I landmark points, which are the demarcation points of the distribution of each zone and can be precisely located; 3, 4, 5, 6, and 7 are defined as type II landmark points, which are the depression and projection points of each part of the wing zone and can be clearly identified; 1, 9, 12, and 14 are defined as type III landmark points, which are the widest, highest, and outermost points of the otolith and are convenient reference points for otolith morphology. TpsDig2 software was used to extract the 14 landmark points for the three geographic populations of jumbo flying squid, and the corresponding x and y coordinate values (2D) and data files were obtained (Fig. 2.23).

Fig. 2.23 Location of the 14 landmark sites for jumbo flying squid statoliths (Chen et al. 2017)

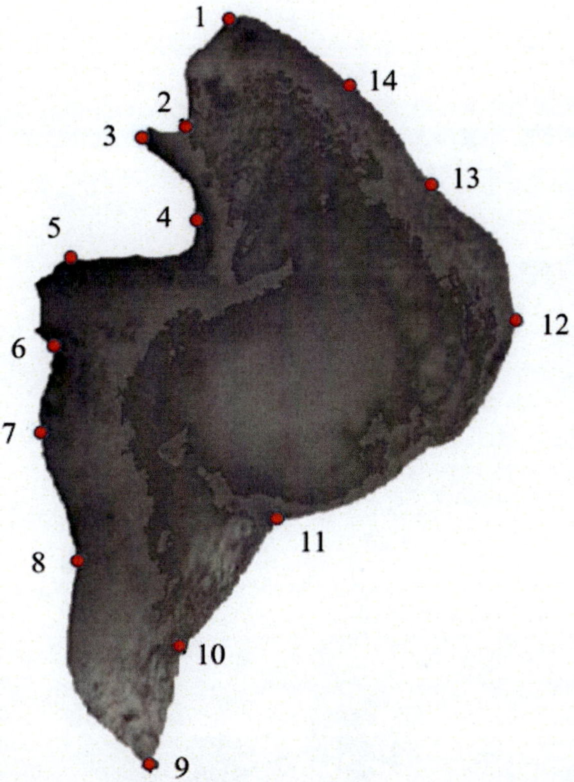

3. Geometric Morphological Analysis
   (A) The validity of landmark points was tested based on least squares regression analysis.
   (B) The landmark points of all samples were superimposed using Procrustes analysis, and the landmark points of each sample were panned, centered, rotated, and scaled using TPSrelw to calculate the centroid size and derive the mean shape, after which the partial warp and relative warp principal component analysis (RCA) was conducted, and the generated analysis report and relative warp (RW) scores were saved.
   (C) Thin-slab sample strip analysis was performed using TPSregr to map the grid deformation of the three geographic populations of stem flexures and to compare and analyze their morphological differences. The TPS series software is available for download at the relevant application website (http://life.bio.sunysb.edu/morph).
   (D) Based on the relative distortion score data for each sample, SPSS 19.0-Bayes discriminant analysis was used to identify the population categories, and the interaction validation method was used to establish the discriminant function and find the discriminant rate.

(2) Stock Identification of Jumbo Flying Squid Based on Landmark Point Analysis

1. Relative Warp Principal Component Analysis

Based on the obtained coordinate point data, the mean shape of the otoliths for the three geographic populations of jumbo flying squid was calculated using TpsRelw software, as shown in Fig. 2.24, and the effect of overlaying landmark points for all samples is shown in Fig. 2.25. The results of the relative warp principal component

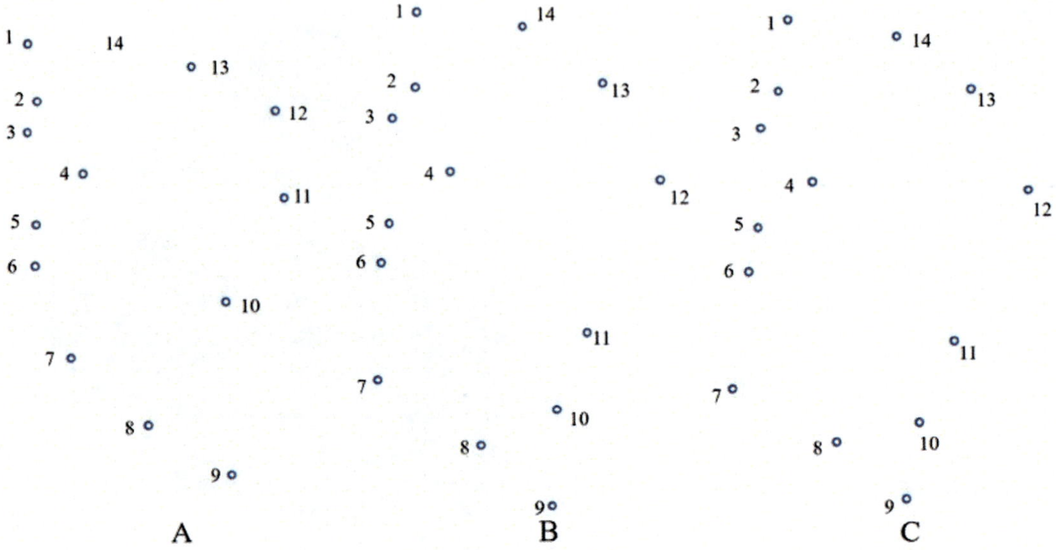

Fig. 2.24 Mean shape of the three local populations of jumbo flying squid (**a**, off Costa Rica; **b**, off Peru; and **c**, off Chile) (Chen et al. 2017)

Fig. 2.25 Overlay of landmark points for the three geographic populations of jumbo flying squid (Chen et al. 2017)

analysis showed that a total of 24 principal components were extracted for the three jumbo flying squid populations, in which the first principal component contribution of jumbo flying squid off Costa Rica was 53.00%, the second principal component contribution reached 15.49%, the third principal component contribution reached 6.27%, and the cumulative contribution of the first three principal components reached 74.76%; the first principal component contribution of jumbo flying squid off Peru reached 21.50%, the second principal component contribution reached 19.66%, the third principal component contribution reached 10.90%, and the cumulative contribution of the first three principal components was 52.05%; the first principal component contribution of jumbo flying squid off Chile was 31.66%, the second principal component contribution was 14.99%, the third principal component contribution was 10.18%, and the cumulative contribution of the first three principal components was 56.84% (Table 2.11). Among the 14 landmark points of the three geographic populations of jumbo flying squid, the cumulative contribution of type I landmark points 2 and 13 and type III landmark point 14 was greatest, mainly in the dorsal region of the otoliths; the

Table 2.11 Eigenvalues and contributions of the top 13 principal components of the relative distortion scores of the three geographic populations of jumbo flying squid (Chen et al. 2017)

Principal component	Eigenvalue			Cumulative contribution rate/%		
	Off Costa Rica	Off Peru	Off Chile	Off Costa Rica	Off Peru	Off Chile
1	1.2503	0.4130	0.5598	53.00	21.50	31.66
2	0.6759	0.3949	0.3852	68.49	41.16	46.66
3	0.4302	0.2940	0.3175	74.76	52.05	56.84
4	0.4082	0.2603	0.2990	80.41	60.59	65.87
5	0.3469	0.2393	0.2465	84.49	67.81	72.01
6	0.2802	0.2253	0.2068	87.15	74.21	76.33
7	0.2657	0.1951	0.1978	89.55	79.01	80.29
8	0.2405	0.1673	0.1877	91.51	82.54	83.85
9	0.2245	0.1489	0.1720	93.22	85.33	86.84
10	0.1849	0.1307	0.1533	94.38	87.49	89.21
11	0.1744	0.1233	0.1259	95.41	89.40	90.81
12	0.1652	0.1178	0.1211	96.33	91.15	92.29
13	0.1520	0.1107	0.1086	97.12	92.70	93.48
14	0.1316	0.1065	0.1035	97.70	94.12	94.56

Table 2.12 Contribution of different landmark sites of the three jumbo flying squid geographic populations at the time of relative warp analysis (Chen et al. 2017)

Landmark	Contribution rate %		
	Off Costa Rica	Off Peru	Off Chile
1	1.58	1.40	2.33
2	32.27	25.76	23.21
3	26.73	22.42	20.17
4	2.51	2.56	4.05
5	15.50	21.84	21.55
6	10.47	15.20	13.34
7	1.54	1.07	1.28
8	1.10	1.74	2.35
9	0.20	0.31	0.70
10	0.44	2.87	4.29
11	1.04	1.48	1.75
12	1.39	0.48	0.67
13	2.28	1.45	2.16
14	2.94	1.46	2.15

contribution of type II landmark points 3, 4, 5, and 6 was greater in the wing region of the otoliths than in the other regions; among the landmark points, type II landmark points played the greatest role in the differentiation of the three geographic populations of jumbo flying squid (Table 2.12).

2. Visual Analysis of Morphological Differences

Relative warp principal component analysis revealed that the morphology of the dorsal and wing regions played a major role in otolith identification for the three geographic populations of jumbo flying squid. Regression score analysis was performed using TPSregr software, and a

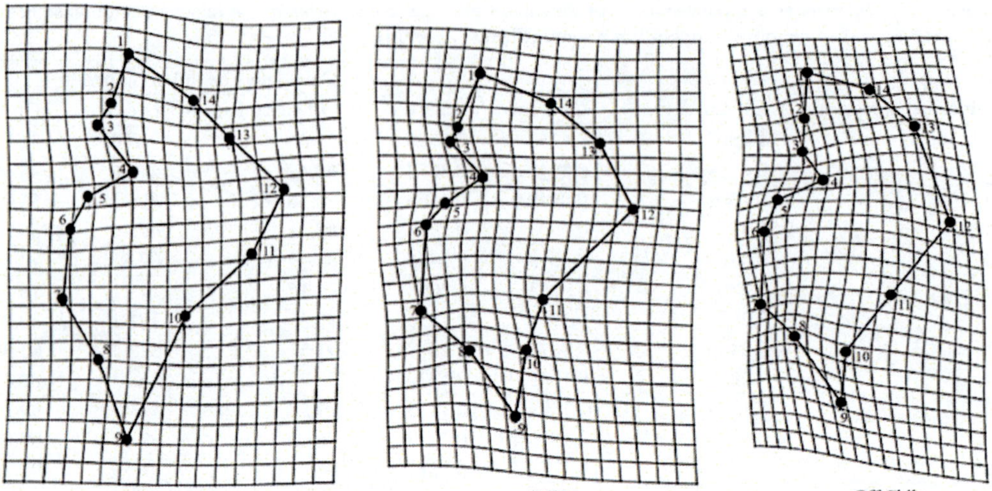

Fig. 2.26 Grid deformation map and visualization of the variation for the three geographic populations of jumbo flying squid (Chen et al. 2017)

Table 2.13 Results of the discrimination analysis of the three geographic populations of jumbo flying squid (Chen et al. 2017)

Step-by-step discriminant analysis	Type	Type: Off Costa Rica	Off Peru	Off Chile	Aggregate	Correctness rate (%)
Initial discriminant	Off Costa Rica	54	0	0	54	100
	Off Peru	0	49	12	61	80.3
	Off Chile	1	12	44	56	78.6
Interactive verification	Off Costa Rica	54	0	0	54	100
	Off Peru	0	47	14	61	77.0
	Off Chile	1	12	44	56	78.6

visual grid map was created by absolute warp (Fig. 2.26). The visualization of the grid warps showed that the three geographic populations of jumbo flying squid, type II landmark points 3, 4, 5, and 6, had the largest warps, with jumbo flying squid landmark point 3 off Chilean waters decreasing relative to that of jumbo flying squid off Costa Rican and Peruvian waters, and landmark point 5 expanded; type I landmark points 2 and 13 had the next largest warps, with jumbo flying squid landmark point 13 off Chile showing an increase relative to the those of the other two areas. The second most warped landmark point is type I landmarks 2 and 13.

3. Discriminant Analysis

The relative warp scores generated by TpsRelw software were used to perform discriminant analysis and cross-validation analysis to establish the corresponding discriminant functions and to determine the discriminant correct rates.

A total of eight variables, PC1, PC11, PC5, PC8, PC12, PC3, PC17, and PC7, were included in the stepwise discriminant analysis using SPSS. The discriminant analysis showed that there were no misclassifications among the 54 jumbo flying squid samples off Costa Rica, with a discriminant rate of 100%; 12 of the 61 jumbo flying squid samples off Peru were misclassified as Chilean samples, with a discriminant rate of 80.3%; and 1 of the 56 Chilean samples was misclassified as Costa Rican and 12 as Peruvian samples, with a discriminant rate of 78.6%. The cross-validation analysis showed the same discrimination rate for jumbo flying squid off Costa Rica and Chile and a slightly lower discrimination rate for jumbo flying squid off Peru, reaching 86.3% and 85.2% for the stepwise discrimination and cross-validation methods, respectively (Table 2.13).

References

Chen XJ (2014) Fisheries resources and fisheries oceanography. Ocean Press, Beijing. (In Chinese)

Chen XJ, Liu BL (2017) Biology of fishery resources. Science Press, Beijing. (In Chinese)

Chen XJ, Lu HJ, Liu BL et al (2011) Age, growth and population structure of jumbo flying squid, *Dosidicus gigas*, based on based on statolith microstructure off the EEZ of Chilean waters. J Mar Biol Assoc U K 91 (1):229–235

Chen XJ, Fang Z, Chen YY et al (2017) Application of geometric morphometry in aquatic organisms. Science Press, Beijing. (In Chinese)

Hammer Ø, Harper DAT, Ryan PD (2001) PAST: paleontological statistics software package for education and data analysis. Palaeontol Electron 4(1):9

Iwata H, Ukai Y (2002) SHAPE: a computer program package for quantitative evaluation of biological shapes based on elliptic Fourier descriptors. J Hered 93(5):384–385

Klingenberg CP (2011) Morpho J: an integrated software package for geometric morphometrics. Mol Ecol Resour 11(2):353–357

Li SF, Wang Q, Chen YL (1986) Biochemical genetic structure and variation of the original population of silver carp, bighead carp and grass carp in the Yangtze, Pearl and Heilongjiang river systems. Aquacult James 10(4):351–372. (In Chinese)

Liu LW (2014) Genetic structure and molecular phylogeography of three oceanic squids. Shanghai Ocean University, Shanghai. (In Chinese)

Liu J, Zhao LJ, Liu QG et al (2015a) Genetic variation of mitochondrial COI genes in natural populations of *Xenocypris argentea* from different aquatic systems. Freshwater Fisheries 45(6):3–8. (In Chinese)

Liu LW, Chen XJ, Xu QH et al (2015b) Analysis of genetic variation between the equatorial population and the Peruvian offshore population of jumbo flying squid based on microsatellite markers. J Ocean Univ China 45(07):53–57. (In Chinese)

Pan Y, Zheng LB, Wang J (2018) Intermediate regional analysis of simple repetitive sequences of five geographic groups of striped zebra sharks. Mar Sci 42 (03):31–36. (In Chinese)

Rohlf FJ (2013) Morphometrics at SUNY Stony Brook [EB/OL]. http://life.bio.sunysb.edu/morph. 16 May 2013

Shan Q, Dong S, Wu HF et al (2006) Restriction fragment length polymorphism analysis of the D-loop region of mtDNA from three populations of bighead carp. Chin Aquatic Sci 2:174–180. (In Chinese)

Sheets HD (2013) Morphometrics software: IMP-Integrated Morphometrics Package [EB/OL]. http://www3.canisius.edu/~sheets/morphsoft.html. 16 May 2013

Thomas F, Jamin E, Wietzerbin K et al (2008) Determination of origin of Atlantic salmon (*Salmo salar*): the use of multiprobe and multielement isotopic analyses in combination with fatty acid composition to assess wild or farmed origin. J Agric Food Chem 56(3):989–997

Yan J, Xu QH, Chen XJ et al (2011) Primary studies on the population genetic structure of *Dosidicus gigas* in the high seas of eastern Pacific Ocean. J Fish CHN 35 (11):1617–1623. (In Chinese)

Ye XC, Chen XJ (2007) Preliminary study on the mantle length composition and sexual maturity of jumbo flying squid off Peruvian waters. J Shanghai Fish Univ 16 (4):347–350. (In Chinese)

Ying YP (2011) Population genetic structure and molecular phylogeography of *Konosirus punctatus* and Japanese sardinella. Ocean University of China, Qingdao. (In Chinese)

Zeng Z, Liu ZJ, Pan LD et al (2012) RAPD analysis of genetic diversity and transformation of SCAR markers in wild population of *Trachidermus fasciatus*. Zool Res 33(02):203–210. (In Chinese)

Check for updates

Life History and Early Development of Fishes

3

Xinjun Chen and Bilin Liu

Abstract

The study of life history of fish is one of the important contents of fishery biology. The life history of fish refers to the life course of individual fish from fertilized eggs to adult fish, and then to aging. The life cycle of fish can be divided into several different developmental stages. Each developmental stage has its own characteristics in morphological structure, ecological habit, and relation with environment. The development process has its own particularity because of the difference of fish species and ecological type. The early stage in the life cycle of fish, that is, from eggs to young fish, is the sensitive period when the fish population is the largest and the death rate is the highest. How much it remains will determine the generation and recruitment of fish. Therefore, it is of great practical significance to study the rule of fish early development in order to clarify the change of fish population and to develop the resource multiplication and protection. This chapter describes in detail the life history of fish and the division and characteristics of the development period, the classification of the life history types of fish, the types of fish eggs, and the morphological characteristics and identification of larvae and juveniles. Moreover, the environmental factors affecting larval survival were also analyzed. The emphasis of this chapter is to master the division and characteristics of different stages of the life cycle of fish and to master the morphological characteristics and identification of eggs, larvae, and juveniles. Mastery of the above knowledge will be of great importance for future research and management in the fields of fishery resources and fishery ecology.

Keywords

Life history of fish · Early development of fish · Egg · Larvae · Juvenile

Abbreviations

PH *pondus hydrogenii*

3.1 Life History of Fish and Their Life Spans

3.1.1 Life History and Division of Developmental Stages

The life history of a fish is its entire life from a fertilized egg to an adult fish to senescence, and this process is also known as the life cycle. The life history of fish can be divided into several

X. Chen (✉) · B. Liu
College of Marine Sciences, Shanghai Ocean University, Shanghai, China
e-mail: xjchen@shou.edu.cn; Bl-liu@shou.edu.cn

X. Chen, B. Liu (eds.), *Biology of Fishery Resources*, https://doi.org/10.1007/978-981-16-6948-4_3

different developmental stages. Each developmental stage has its own characteristics in terms of morphology, ecological habits, and connection with the environment. The developmental process of fish differs depending on the fish species and ecological type.

The life history process and its developmental stages are described as follows, using the vast majority of oviparous osteichthyes as examples (Yin 1995):

1. Embryo stage. This is the period when individual fish develop within the egg membrane. The embryo stage begins when the sperm enters the pore of the egg membrane and the sperm-egg union process is completed. This period is characterized by the development of the embryo being confined to the egg membrane, and therefore, it is also called the egg development stage. The embryo is entirely dependent on the yolk for its nutrition and is associated with the environment, mainly with respiration and predation by enemies.
2. Larval stage. This is the period when the fry hatch out of the membrane and change from developing inside the egg membrane to developing outside the egg membrane, before the mouth is opened, and this stage is endogenous in nature (relying on yolk and oil globules) and changes based on the internal environment of the parent to developing directly in the external environment. When the embryo hatches out of the membrane, it enters the pup stage. The ex-hatchling is transparent, blood is often unpigmented, eye pigment is partially formed or unformed, each fin is membranous and finless, the mouth and digestive tract are incompletely developed, and there is a large yolk sac as a source of nutrition. This stage is also known as the yolk sac stage pups, or prelarva. Unlike the embryonic stage, which is still dominated by respiration and defense against predation by enemies, the yolk sac stage pups begin to acquire the ability to avoid enemies and behavioral characteristics. Thereafter, as the young develop further, eye, fin, mouth, and digestive tract functions are gradually formed; gill development begins; roving patterns are established; and the young begin to obtain food from the outside world. At this stage, young fish are generally pelagic in the same water layer, dissolved oxygen conditions improve, and contact with the outside world gradually shifts to nutrition and protection from enemies.
3. Juvenile stage. This is the period when the body shape rapidly approaches that of an adult fish. The juvenile stage is marked by the disappearance of juvenile characteristics such as transparency, the initial formation of fins, and, in particular, the beginning of the scale formation process. Early juvenile fish generally still live a planktonic life and only at a later stage do they shift to the inherent lifestyle of each group. Contact with the outside world during this period is mainly for nutrition purposes and defense against enemies. At this stage, the digestive organs develop qualitatively into the basic types of adult fish, and the stomach, intestine, pyloric pendulum, etc. reach the type and number inherent to each "species." The complete development of the scutum and the completion of metamorphosis are the signs that this stage is ending. A major feature of the ecological habits of this period is the significant increase in clustering.
4. Young stage. This stage represents the fastest growing period in the life of an individual, and the appearance of a fish at this stage is identical to adult fish in body shape; however, the markings and coloration are still in the process of change. At this stage, all scales are formed, the fins and lateral lines are fully developed, and body color, markings, proportion of various parts of the body, and habitat habits are consistent with those of adults, and at this point, the fish enter the juvenile stage. The few species that are ovoviviparous or fecund are often produced from the mother as juveniles. During the juvenile stage, the gonads are not yet developed, and secondary sexual characteristics are not obvious or are absent. This period is usually a period of rapid growth, and as the fish grow rapidly,

the adaptive relationship with the external world becomes increasingly weaker in terms of defending against enemies, natural mortality decreases, and nutritional relationships become increasingly important.

5. Immature stage. This is the period when the morphology is identical to that of the adult fish but the gonads have not yet matured, and this is generally the period of transition from juvenile fish to sexually mature fish.
6. Mature stage. The adult stage begins when the gonads first mature. Mature individuals are able to reproduce in the appropriate season and reproduce offspring; if there are secondary sexual characteristics, they have already appeared at this time. Some large and medium fish that reach sexual maturity late do not yet have mature gonads when they reach food size, so they are called food fish. Another extremely important relationship between this stage and the outside world, in addition to nutrition, is reproduction. Most of the nutrients consumed by an individual are used for gonad development and accumulation of reserves of fat and other substances for offspring needs during migration, overwintering, and reproduction. Natural mortality is minimized, while fishing mortality increases sharply. Reproductive capacity develops, and secondary sexual characteristics develop during a certain season of the year when reproductive development takes place.
7. Aging stage. This stage has no clear boundaries. It is generally the period when sexual function begins to decline, fertility decreases significantly, and growth in terms of length is extremely slow. The nutrients consumed by a fish are mainly used to sustain life and accumulate energy substances such as fat to maintain metabolic activities in case of an urgent need. In unfished waters, natural mortality rates begin to rise again.

Studies have concluded that individual development in fish, in general, proceeds in a continuous and progressive manner. However, the transition from one developmental stage to another is often accomplished in an abrupt manner within a fairly short period of time. This means that at the various developmental stages, fish undergo slow and gradual changes, such as the accumulation of material, without experiencing essential changes in morphology, ecology, and physiology. When this gradual change has reached a certain point, a fish body often completes a sudden change to another developmental stage within a short period of time, sometimes in just a few hours, when the morphology, ecology, and physiology of the fish body have all experience essential changes. Thus, fish of the same species at different developmental stages maintain a certain degree of independence in their morphology, ecology, and physiology, as well as in the way they relate to the external environment, and this independence is expressed in very different ways among different species and ecological types of fish.

3.1.2 Types of Fish Life Histories

During the long evolution of fish, the specific environment in which they live and their inherent morphological, physiological, and ecological characteristics have led to differences in the lengths of the life cycles of various species of fish. Some sturgeon species are known to live for hundreds of years, while small tropical fish have a shorter life span; some gobies even live for only a few months. The life cycles of different populations of the same species also tend to vary markedly; for example, the life cycles of the Daiqu, Min-Yuedong, and Naozhou groups on the Chinese coast are approximately 30 years, 12 years, and 9 years, respectively.

In general, the life cycle of fish lengthens with increasing geographical latitude; i.e., fish living in tropical low-latitude waters have a shorter life cycle than those living in midlatitude and high-latitude waters. Since there are more significant differences in ecological habits between fish with long life cycles and those with short life cycles, the life cycles of fish have been further classified into three different types based on studies (Chen 2014; Chen and Liu 2017).

1. Single-cycle fish. Single-cycle fish are sexually mature at 1 week of age, reproduce only once in their lifetime, die after giving birth, and consist of only one age class; examples of single-cycle fish are silverfish (*Protosalanx hyalocranius*), yellowfin goby (*Acanthogobius hasta*), etc. For single-cycle fish, the reproductive population consists entirely of recruitment, and essentially all individuals who participate in reproductive activities die afterwards. Therefore, the number of recruitments each year determines the number of reproductive groups, and the abundance and failure of generations profoundly affect the population size. Therefore, this type of fish has more dramatic population size changes, and its variability is large. Intensively fished, the stock is vulnerable not only to destruction but also to recovery.
2. Short-cycle fish. Although short-cycle fish can sexually mature repeatedly, they have a short life span and simple age groups, and examples of short-cycle fish include Japanese scad, sardines, anchovies, and some common small fish. However, the population structure and timing of sexual maturity vary considerably among species. The variability in their numbers is often large, which also means that fish resources are vulnerable to damage, such as overfishing. However, with adequate management measures, stocks can be easily restored.
3. Long-cycle fish. Some large and medium carnivorous fishes, such as large yellow croaker and bastard halibut, have a long life cycle, a large number of repeated spawning events during their lifetimes, and a complex age structure, and their resources change relatively smoothly from year to year, with a moderate process of change and a small range of change; however, the rate of recovery is also slow after the destruction of this type of fish resource.

The main significance of the different life histories of fish, which are inherent biological characteristics of individual species, is that they serve as basic information for studying population characteristics and determining the number of fish in a population and their dynamics of change. With the development of and improvements in fisheries resource science, research in this area has become even more important, and a large number of research results have been obtained.

3.1.3 Longevity

The durations of the early stages of individual fish development are usually much shorter than the durations of the later stages. Early developmental stages are usually completed in a few days to a few months, while later developmental stages are related to longevity. Life span is the amount of time that a fish lives, and it depends on the genetic characteristics of the fish and the external environmental conditions where it lives. In nature, only a very small number of adult fish produce offspring that can complete their entire life history and live their physiological life span, while the vast majority of fish cannot complete their entire life history due to unsuitable external environmental conditions. The life span of a fish is called its ecological longevity.

Various fish species have different life spans, and their individual sizes vary. Generally, long-lived fish are large individuals; short-lived fish are small individuals. The differences in the life spans and maximum sizes of different fish are very large. The largest known fish in the world is the cartilaginous fish living in the ocean, the whale shark (*Rhincodon typus*), which can reach 18–20 m in length and weigh over 10 tons, and its life span is unknown. Among bony fishes, species of sturgeon (Acipenseridae) and paddlefish (Polyodontidae) experience longevity and are large in size. The European sturgeon of the Caspian and Black Seas are reported to reach 9 m in length, weigh approximately 1.5 tons, and live to be >100 years old; Chinese paddlefish (*Psephurus gladius*) in China is the largest freshwater fish in the world, with a maximum individual size of 7 m, a weight of over 1 ton, and a typical life span of 20–30 years, with a maximum life span of over 100 years. However, some fish of the family Gobiidae and Scopelidae, Japanese killifish, and silverfish only live for 1 year or less. Most small and medium cephalopods also only live for 1 year or less. These species generally die after reproduction. The smallest known fish in the world is the goby (*Trimmatom nanus*) living in the Philippines, with sexually mature individuals measuring only 0.75–1.15 cm (Chen 2014; Chen and Liu 2017).

Although the lengths of the life spans of fish vary greatly between species, the vast majority of fish live between 2 and 20 years, with approximately 60 percent of them living between 5 and 20 years, no more than 10 percent living for more than 30 years, and approximately 5 percent living less than 2 years. There are many species of Chinese freshwater fish that live 2–4 years; anadromous salmon generally live 3–6 years; many medium and large freshwater fishes, such as black carp, grass carp, silver carp, and bighead carp, generally live 7–8 years. Few fish live more than 10 years old, but some can live up to 15–20 years. In comparison to other fish species, sturgeon fish species have longer life spans, generally reaching 20–30 years of age, with sexual maturity occurring only at 10 years of age or older. Marine fishes have shorter life spans, such as anchovies that generally only live to 3 years old, but large yellow croaker *Larimichthys crocea* off China can live up to 29 years old (Chen 2014; Chen and Liu 2017).

Different geographical populations of the same species of fish have different life spans. For example, the maximum life span of the large yellow croaker inhabiting the coast of Zhejiang, China, is 29 years; those living off the coast of Fujian and Guangdong can live up to 17 years, while those living in the eastern waters of Hainan Island live only to 9 years old. This difference is the result of the influence of different living environments on the life cycles of the populations. Another example is the maximum body length of herring populations in Icelandic and Norwegian sea areas that can reach 37–38 cm and live to 22–23 years old, while herring living in the English Channel, North Sea, and Baltic Sea have a maximum body length of 20–32 cm and a maximum age of 10–13 years old; Kuril Islands herring length and age are between those of the above two populations, with a maximum body length of approximately 35 cm and a maximum age of 15–17 (Chen 2014; Chen and Liu 2017).

3.2 Early Development of Fish

3.2.1 General Characteristics and Processes of Early Fish Development

The early stages of the fish life cycle, from egg to juvenile, are sensitive periods when fish numbers are greatest and mortality is highest, that is, when the rate of change in fish numbers is highest. The amount of its residuals will determine the amount of occurrence and replenishment of fish generations. Therefore, it is of great practical importance to conduct research on the early developmental patterns of fish to elucidate the changes in fish populations and to carry out resource enhancement and conservation. Taking dotted gizzard shad *Konosirus punctatus* as an example, we briefly describe the main morphological characteristics and ecological habits during its early development (Fig. 3.1).

Dotted gizzard shad is a species of herrings with a long, oval, laterally flattened body that is approximately 20 cm long and is found from India to the East Indies and in Korea and southern Japan. Dotted gizzard shad is distributed along the coast of China and is a pelagic fish in China's coastal waters, preferring to inhabit coastal harbors and estuaries at depths of 5–15 m. It can live in both seawater and salty water and sometimes can enter freshwater without dying. It spawns offshore and in estuaries and feeds on phytoplankton and zooplankton, many algae species, shellfish, crustaceans and copepod larvae, foraminifera, sand shell ciliates, etc. Sometimes, it also feeds on benthic organisms, plankton, and small crustaceans.

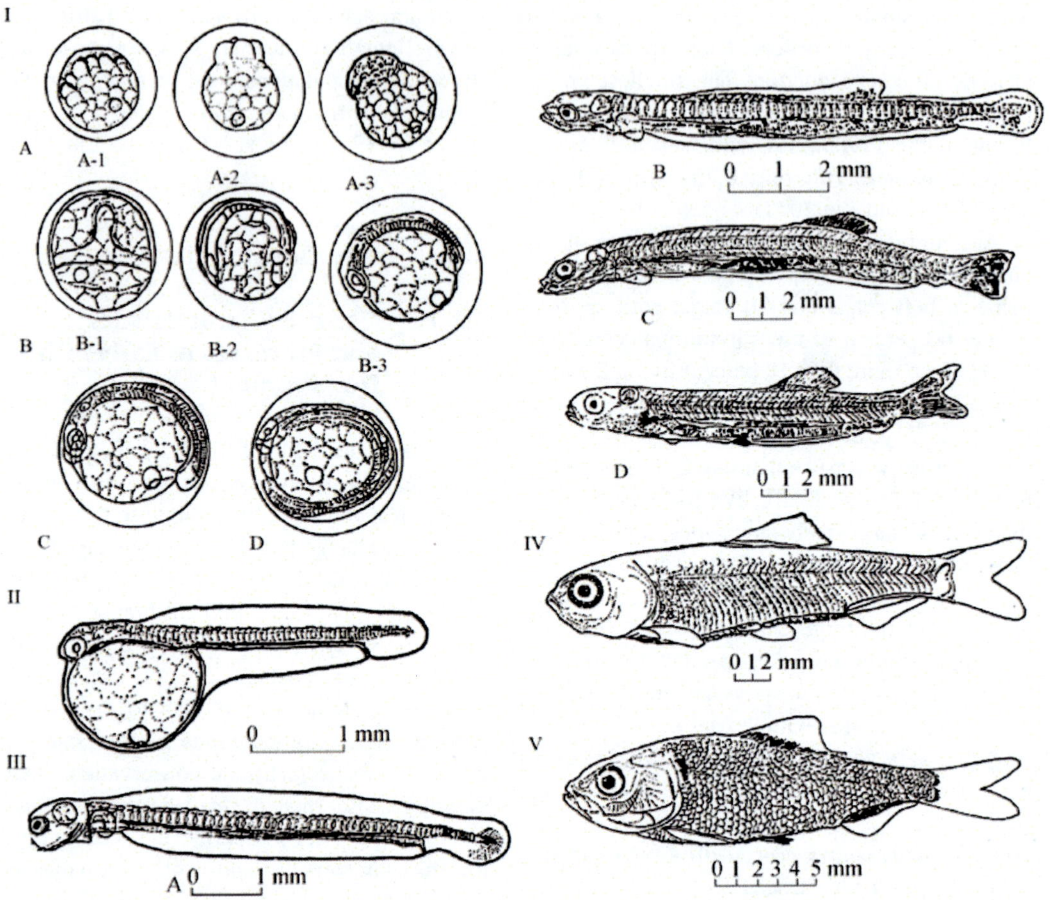

Fig. 3.1 Map of early development of dotted gizzard shad in the Yellow Sea and Bohai Sea (Chen 1997)
Stage I – intraovarian development: *A* oogenesis stage, *A-1* unicellular phase, *A-2* quadruplet phase, *A-3* early embryonic phase, *B* protointestine formation stage, *B-1* postembryonic phase, *B-2* optic vesicle phase, *B-3* phase of closure of embryo hole, *C* embryo formation stage, *D* hatching stage. Stage II – early juvenile. Stage III – late juvenile: *A* oocyst disappearance stage, *B* dorsal fin bar emergence stage, *C* anal fin bar emergence stage, and *D* ventral fin bar emergence stage. Stage IV – juvenile stage. Stage V – young stage

(1) Egg stage. The gizzard shad had round, floating eggs with smooth, colorless, and transparent membranes, with a diameter of 1.15–1.55 mm. The eggs are yellowish and transparent, covered with reticulated textures and containing a light brownish red oil ball, with a yolk diameter of 0.8–1.0 mm. The development of fertilized eggs can be divided into the following four different developmental stages (developmental water temperature of 15.5–18.0 °C):

1. Oogenesis stage. The period from the fertilized egg to the end of the blastocyst stage. It is characterized by one single-celled organism that undergoes frequent divisions to become a multicellular body at the blastocyst stage. This stage lasts approximately 10–12 h, with the one- to four-celled early segmented embryo stage lasting approximately 3–4 h.
2. Protoderm formation stage. This stage starts from the beginning of protointestinal action until the closure of the embryonic pore. This stage continues for the longest time and has more complex changes, such as mesodermal differentiation and the

appearance of neuroectoderm and myotomes, and it takes approximately 18–22 h to complete this stage.

3. Embryo formation stage. This stage occurs from the formation of the tail bud to the beginning of the heart beating. This stage is characterized by the separation of the progenitors of the major organ systems and the differentiation of tissue structures and the beginning of the appearance of pigment cells. The duration is approximately 13–16 h.
4. Incubation stage. This stage occurs from the time the heart starts beating to the time the embryo breaks the membrane and hatches. The duration is approximately about 6–8 h. It is characterized by significant growth of the embryo in terms of length, with accelerated growth of the tail bud in particular; separation of the head from the yolk; the emergence of hatching glands; and the constant contraction and oscillation of the embryo body within the egg membrane, which is about to break the membrane for hatching. At this point, the embryonic stage ends, and the duration of this stage is 51–57 h.

(2) Juvenile stage. The whole length of the first hatchling of dotted gizzard shad is only 0.4–4.0 mm and 52 (44 + 8) muscular segments. There is a well-developed yolk sac that is transparent and sparsely pigmented. The young fish float on the surface and swim in waves occasionally. When the water temperature is 18–22.6 °C, the fish body appears with the pectoral fin bud, melanin increases, and the yolk sac is gradually absorbed. By the time the mouth is first opened, the intestinal canal is obvious, and after 3 days, the full length of the fish is 5.5–6 mm, thus ending the juvenile period. The fry also swim in the water with a supine to vertical posture to vertical upside down, and finally, they began to turn flat, entering the late juvenile stage.

(3) Late broodstock stage. From this time onwards, the fry are transformed into a period of exogenous nutrition. Because of the complex metamorphosis of the fry and its long duration, this stage can be divided into the following four specific stages (developmental water temperature of 18.0–28.20 °C):

1. Yolk sac disappearance stage. This stage involves 4–5 days into hatching with the full length of this stage at 5.5–6.5 mm and 55 (43 + 12) muscular segments. At this time, the fish is elongated, the mouth is obvious, the intestinal lumen folds are visible, and the young fish turns to wavelike flat swimming. Ingestion of tiny organisms such as yeast begins.
2. Dorsal caudal fin primordia and fin emergence stage. This stage is entered 6–10 days after hatching. The fish are at full length at 6.5–8.5 mm or 10–11 mm (post), with 54 (43 + 11) muscular segments. The lower jaw exceeds the upper jaw in length at this time. The auditory capsule reaches its maximum, the anal and caudal fins each have a tuft of melanophores, and the dorsal and caudal fin primordia appear. By the time the dorsal and caudal fin bars emerge, two clumps of melanin also appear on the posterior edge of the head. The fry are feeding on bivalve, such as trochophore.
3. Anal fin primordium and fin emergence stage. This stage is entered 12–16 days after hatching. The full length at this stage is 10.5–11 mm (pre) or 11–14 mm (post), with 52 (42 + 10) muscular segments. When the anal fin primordium appears, the melanin of the fish increases, but except for the deepening of the anal pigment tufts, pigment bands have not yet formed at the upper and lower margins of the intestinal canal. Twelve dorsal fin strips and 12 anal fin strips appear, the pigment at the auditory capsule has become a "\ /" type pigment tuft, and the caudal fin also begins to divide into upper and lower lobes. At this stage, bivalve larvae, marine copepods, and their larvae are consumed in large numbers.
4. Ventral fin bud and ventral fin stripe emergence stage. This stage is entered 18–23 days after hatching. The full length

at this stage is 15.5–17 mm (pre) or 17–20 mm (post), with 50 (41 + 9) muscular segments. When ventral fin buds appear, 15–16 dorsal fins and 13–15 anal fin bars appear, and yellow pigment spots appear on the mid-axis part of the fish. With the appearance of ventral fin bars, the mid-axis yellow pigment spots gradually become yellow pigment bands, and the swim bladder starts to inflate. The number of odd fins also stabilizes (16–18 dorsal fins and 12–19 anal fins). At this stage, in addition to feeding on copepods, the fry also feed heavily on haloarchaeal larvae during artificial breeding, which results in strong feeding and rapid growth.

(4) Juvenile fish stage. The fry enter this developmental stage approximately 28 days after hatching, and the water temperature is 23–25.0 °C. The body height of the 22 mm fry increases significantly, and the odd and even fins resemble those of adult fish, reaching a fixed number (dorsal fins 16–17, anal fins 22–24, pectoral fins 15, and ventral fins 7). The prismatic scales appear, and the body is gradually opaque, which is the stage of juvenile fish. When the full length reaches 30 mm, the body length reaches 25 mm, the muscular segments reach 19 + 14, and juvenile development is complete. When scales begin to appear and the body is greenish green on the back and silvery white on the ventral side, the fish has reached the late juvenile stage. At this time, the fry are large and active, and they begin to feed in groups. In addition to swallowing copepods in large quantities, they also strongly feed on artificially fed brine shrimp larvae.

(5) Young stage. After approximately 35 days, the fry reached 36 mm in length and 30 mm in body length. Scales cover the body surface, the fins at the posterior end of the dorsal fin begin to lengthen, and the black spot above the posterior gill cover is not yet obvious; that is, they have entered the young stage. This is the end of the early development of dotted gizzard shad.

The morphological and ecological characteristics of the early development of fish vary from species to species, especially the descending riverine eel, the deep-sea monkfish, and the bottom-dwelling flounder and other fishes, whose early developmental morphology differs greatly from that of the dotted gizzard shad, but they usually go through the abovementioned major developmental stages and have the abovementioned basic characteristics. Therefore, it is possible to identify a wide variety of young and juvenile fish in the ocean by following their patterns and observing them carefully.

3.2.2 Implications for Research on Early Development in Fish

Synthesizing the state of national and international research, the significance of research on eggs and young individuals can be summarized into the following four areas:

1. Eggs and young fish are used as the object of study to learn about embryonic development and the morphology and classification of juvenile fish and their growth, mortality, physiology, and ecological habits. The reason for this focus of study is that an egg, whether it is floating, sinking, or adhering, and its development and hatching, from an almost passive yolk sac-bearing young fish to the passive drifting prefish that depends on the yolk sac for nutrition to a later young fish that can swim freely and can actively suck and feed, and even further to a young fish that can swim on the surface of the water body or reside on the bottom, have several morphological, physiological, and developmental stages with different characteristics such as their ecology.
2. The ecology of marine waters (or freshwater) is studied in terms of studying eggs and juvenile fishes as predators and indicators to evaluate the role of pollution.
3. As a breeding object, in addition to studying fry needs related to aquaculture, it is also

necessary to study the selection of eggs and young of good species.

4. In terms of natural resource recruitment or fishery forecasts, studying the growth and survival numbers of eggs and juvenile fishes provides the basic data for measuring the size of parental resources and forecasting the recruitment, so spawning ground surveys are important information for conducting accurate resource forecasts and analyses.

To integrate morphology, function, and environment, the theory of morphological developmental stages of fish can be summarized as follows: (1) the individual development process of fish can be divided into many small developmental stages according to their morphological characteristics; (2) within one developmental stage (based on a certain body length change), there is usually only quantitative growth and no qualitative changes in morphology and ecology, but when moving to the next stage, almost all organ systems grow; and (3) in each developmental stage, fish have a special relationship with their environment, and their morphological characteristics are the result adapting to the environment.

Therefore, it is of great importance to conduct in-depth research on the life cycles of fish, to determine the interrelationship between each developmental stage and the environment and to elucidate the basic laws of their life activities for exploiting fishery resources, scientifically managing, and stocking.

3.3 Morphology and Identification of Eggs and Juvenile Fish

3.3.1 Morphological Structure and Identification Points of Fish Eggs

(1) Morphological Structure of Fish Eggs

The egg is a highly specialized cell with specific adaptations for fertilization, embryonic development, and nutrition, and its structure consists of the following components:

1. Ovum. The membranes are located in the outermost layer of the egg, protecting the oocyte from external factors and maintaining the egg in a certain shape, acting as a barrier to the external environment to ensure proper development of the embryo. The thickness and configuration of the egg membrane vary depending on the species and the conditions under which the cells mature.

The surface of the egg membrane is generally smooth, clear, and hornlike, but some species have special structures on the egg membrane, such as Rajiformes (skates, etc.), where the ovoid shape is very large and covered with a horny eggshell. The largest eggshells are 180 mm long and 140 mm wide and have a boxy, spiral shape, often with curling filaments wrapped around the outside of the eggshell on algae or rocks to provide a stable incubation environment. Lizard fishes have wrinkled or irregularly fragmented egg membranes. The egg membranes of deep-sea luminous fishes have many triangular columnar protrusions, and the protrusions vary by species. The eggs of *Cypselurus agoo* are sticky, with thick egg membranes and 30 to 50 rice-like strips on the surface, by which the eggs attach to the seaweed (Figs. 3.2 and 3.3). The egg membranes of *Trichiurus lepturus* are pale red. The egg membranes of Japanese sardinella are slightly light blue.

2. Yolk. A yolk is a special protein that is formed from the vesicles of egg cytoplasm and is a nutrient required for embryonic development. The size of a yolk is generally related to the length of time an embryo has been developing. Embryos with large yolks take longer to develop, while those with small yolks take less time to develop.

A yolk can be various colors, ranging from light red to pale green, but the vast majority of yolks are yellow and transparent and opaque. The shape of a yolk varies with the amount of yolk and is often finely granular in eggs that are not

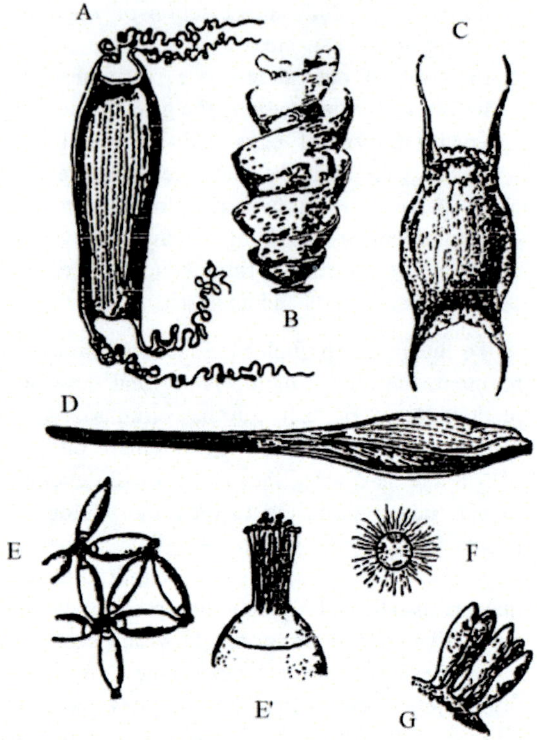

Fig. 3.2 Morphology of marine fish eggs (Chen 1997)
A Brownbanded bamboo shark, *B* tiger shark, *C* yellowing flying fish, *D* eggshell of ratfish, *E* egg of Pacific hagfish, *E* animal pole of one egg, *F* needlefishes, and *G* black goby

very rich in yolk and spherical in eggs that have many yolks and a large egg mass. The amount of yolk content and its distribution determine the manner of subsequent oogenesis and the size of the divisions. Depending on the amount of yolk and the location of the yolk distribution, the eggs can be divided into four types: even-yolked, inter-yolked, medium-yolked, and telangiectatic eggs.

The surface structure of a yolk varies according to the species; some are uniform; others have a cracked surface, e.g., irregular reticulated cracks on the yolk surface of dotted gizzard shad. The surface of the yolk of Carangidae is neatly cracked in the form of vesicles, and the yolk of the milkfish *Chanos chanos* has small, finely arranged dots.

3. Oil globules. A special component of the eggs of many species of Scleractinian fishes is the small globular body containing fat that is surrounded by a protoplasmic film, which not only is a nutrient store for floating eggs but also acts as a "float" to keep the eggs in a certain layer of water; however, it is only a nutrient store for sinking eggs.

Generally, the oil globules are round and spherical, but some species have deformed oil globules that occur during development. Some eggs contain only one oil globule (e.g., mackerel), while others contain multiple oil globules of different sizes (e.g., shad and *Coilia*) or more and smaller oil globules (e.g., puffer, *Tetraodon fluviatilis*).

Some species have no oil globules, although they are floating eggs, such as those of the lizard fish and *Fistularia villosa*. In addition to the differences in the number of oil globules between various species, the color of the oil globules varies, from pale yellow to dark green and orange, but they are generally very transparent.

4. Ooplasm. The ooplasm is the cytoplasm (protoplasm) of the egg, which is the main part of the oocyte body and is the center of nutrition and vital activity of the oocyte. The amount of

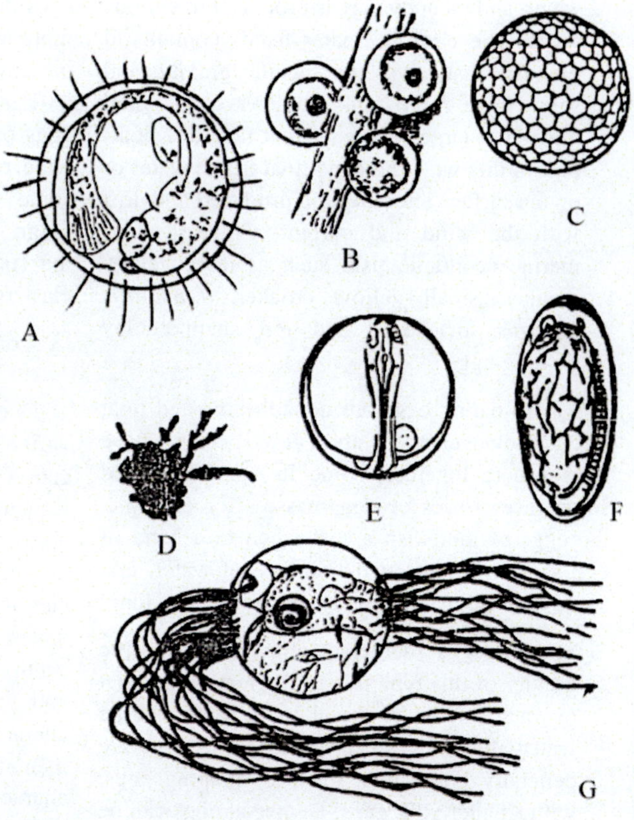

Fig. 3.3 Morphology of marine fish eggs (Chen 1997) *1* flying fish eggs, *2* pufferfish eggs; *3* striped sole eggs, 4. greenlings eggs attached to seaweed, *5* silver croaker eggs, *6* anchovy eggs, and *7* *Cypselurus agoo* eggs

cytoplasm within a fish egg determines the size of the cell.

5. Egg nucleus. Additionally, as the reproductive nucleus or nucleus, the nucleus is directly related to egg cleavage, growth, and metabolism. The nucleus is usually round or rod-shaped and relatively large, and its location is not visible under normal conditions, sometimes on the lateral side of the egg, sometimes in the middle, but usually on the more polar side where the cytoplasm is abundant.
6. Polarity. The polarity of an egg is due to the uneven distribution of yolk in the ooplasm (cytoplasm). The end with more yolk is called the vegetative pole, and the end with less yolk or no yolk (i.e., the end where the cytoplasm is mainly concentrated) is called the animal pole. When resting, the animal pole always faces down, and the vegetative pole faces up. A fertilized egg forms the germinal disc at the animal pole, and cell division begins at the germinal disc, at which point the position of the animal pole is more easily seen.
7. Yolk gap or perivitelline cavity. This refers to the space between the oolemma and the oocyte proper. The perivitelline cavity of a fertilized egg will increase in size as the sperm enters and swells with water absorption.

(2) Types of Fish Eggs

Fish egg types can generally be divided into two main categories according to ecology and morphology. The following two types of eggs can be distinguished on the basis of their different specific gravities and properties, the presence or absence of adhesion, and the strength of adhesion:

1. Floating eggs (pelagic eggs). The specific gravity of this type of egg is less than that of

water, and its buoyancy is produced in various ways. The eggs of many fishes contain oil globules that lower the specific gravity, and some have large egg diameters and small grains but large yolk gaps that facilitate floating. In this way, the eggs float in the water or on the surface after they are produced and drift with the wind and current. China's major marine economic fish, such as large yellow croaker, small yellow croaker, *Trichiurus lepturus*, mackerel, and red snapper, lay floating eggs.

Most floating eggs are nonadhesive and float freely. However, there are a few species whose eggs adhere together, some in egg bands and some in egg sacs or egg masses; for example, the eggs of monkfish are attached to a band of egg sacs and float on the surface of water, and some of these masses can be several meters long.

2. Sinking eggs (demersal eggs). The specific gravity of this type of egg is greater than that of water, and the eggs sink to the bottom of the water after they are produced. The eggs are generally larger than those of floating eggs, with smaller yolk gaps. Demersal eggs can be subdivided into (1) nonattached demersal eggs, the eggs sink to the bottom or in pits dug by the parent fish and are not attached to objects, and (2) attached demersal eggs, of the attached type, there are two types: adherent and attached. The egg membrane, which is attached to the egg, has its own mucus and is attached to other objects; the attached egg has an attachment on it and is fixed to other objects by the attachment. (3) Filamentous twining eggs. These eggs are spherical, without oil globules, with a thick egg membrane and 30–50 filaments on the surface, which are approximately five to ten times as long as the diameter of the egg, and they are distributed at the two poles of the egg membrane, by which the eggs attach to the seaweed. The number of fish with sinker eggs is small.

Some fish eggs have intermediate characteristics between those listed two egg types, with slightly sticky egg membranes. For pikes living in brackish water and freshwater, eggs float in seawater with a salinity of 0.015 or more, are suspended in the middle layer of water in semisaline water with a salinity of 0.008 to 0.01, and sink to the bottom in freshwater. The eggs of other fish species are distributed over a wide range of depths; for example, the eggs of some species of the cod can be trawled in the range of 1000–2000 m in the deep sea and in the 100 m deep sea, which makes it difficult to classify them.

(3) Points for Identification of Fish Eggs

Due to the diversity of fish species and their variability during early development, the identification of fish species can be difficult, and it is difficult to find a systematic and practical searchable list of eggs and young fish. To identify eggs, the first step is to know and understand the species and their spawning period, the area of the sea where it occurs, and the season during which it occurs to determine the possible species of eggs and the "stable" morphological and ecological characteristics of the eggs at different developmental stages, especially the external characteristics of the eggs. Other identification characteristics are briefly listed below:

1. Types of fish eggs. Eggs can float (free eggs, e.g., gizzard shad, or cohesive eggs, e.g., monkfish) or sink (attached eggs, e.g., *Hemiramphus far*, or nonadherent eggs, e.g., salmon and trout).
2. Egg size and shape. Egg diameter size and shape are one of the main bases for identifying fish species, such as anchovy and goby whose eggs are oval; however, the former has the free-type floating eggs, while the latter has sunken eggs with fixed filaments that attach to the spawning chamber on the wall of the chamber hole (*Acanthogobius hasta*) or empty shells (*Tridentiger trigonocephalus*). For example, the species in the Yellow and Bohai Sea have the same round floating eggs; however, *Trichiurus lepturus* eggs have diameters of 1.79–2.20 mm, and small yellow croaker in the Bohai Sea eggs have diameter of 1–1.65 mm.
3. Egg membrane characteristics. The egg membranes of marine fish are usually thin,

smooth, and transparent. However, some species have hexagonal fissures and netlike patterns on their egg membranes (striped sole); some have small spinelike protrusions on their egg membranes; others have finer filaments on the surface of their egg membranes (*Cypselurus agoo*, large silverfish, etc.).

4. Yolk structure. The structure and morphology of yolks vary depending on the abundance of yolk content; e.g., most floating eggs have evenly distributed, transparent, slightly yellow yolks, but the dotted gizzard shad, etc. have yolks with irregular reticulate textures due to coarser yolk grains.
5. Oil globules. The presence or absence of oil globules within eggs and their number, size, color, and distribution are important in identifying eggs; e.g., bastard halibut has only one large oil globule. Striped sole, on the other hand, has dozens of small oil globules.
6. Yolk gap. The size of a yolk gap (perivitelline cavity) varies among fish of the same species or different species.
7. Embryo characteristics. After the formation of an embryo, which is the more "stable" stage of the egg's external morphology throughout its development, a very important period for identifying an egg occurs because the shape and size of the embryo body and the early and late appearance, shape, and distribution of pigmentation are the most important basis for egg identification.

3.3.2 Young and Juvenile Fish and Their Identification Points

The main points and methods for identifying young and juvenile fish are the same as those for eggs. Knowledge of the shape and characteristics of fry at each stage of development is the basis for identifying young and juvenile fish.

1. Young fish stage. The shape of the fish, the shape of the yolk sac, the position of the oil bulb in the sac, the position of the anus, the shape of the fin membrane, the number of myotomes, and the shape, color, and distribution of pigmentation are the main characteristics for identifying species of young fish.
2. Late juvenile fish. The length of the fish, the ratio of body length to each part, the position of the anal opening, the number of myotomes, the type and arrangement of pigmentation, and the shape and position of each fin primordium or fin strip are the main characteristics for identifying species of late juvenile fish.
3. Juvenile stage. In addition to the same points of identification for the late juveniles, more attention should be given to countable traits and measurements such as the shape of the head and tail, the number of fins, and the number of vertebrae (Fig. 3.4).

The three most common methods currently used nationally and internationally in studying morphological changes in marine fishes at various stages of development are briefly described as follows:

1. Artificial intelligence method. Information obtained on an adult fish by artificial intelligence can be compared with naturally collected samples and used to identify the species. Species identification with this method is reliable.
2. Dynamic study method. A large number of specimens of different sizes are used to follow the developmental stages in sequence, to compare morphology and organ development, and to classify them according to morphological aspects.
3. Static research method. This research method focuses on the integrity of a single individual and traces the main features of the morphological developmental stage of an individual. It has the advantage that even if only a few specimens are available, they can be identified and classified. However, the use of this method requires familiarity with the morphological characteristics of juveniles of various families, genera, and species.

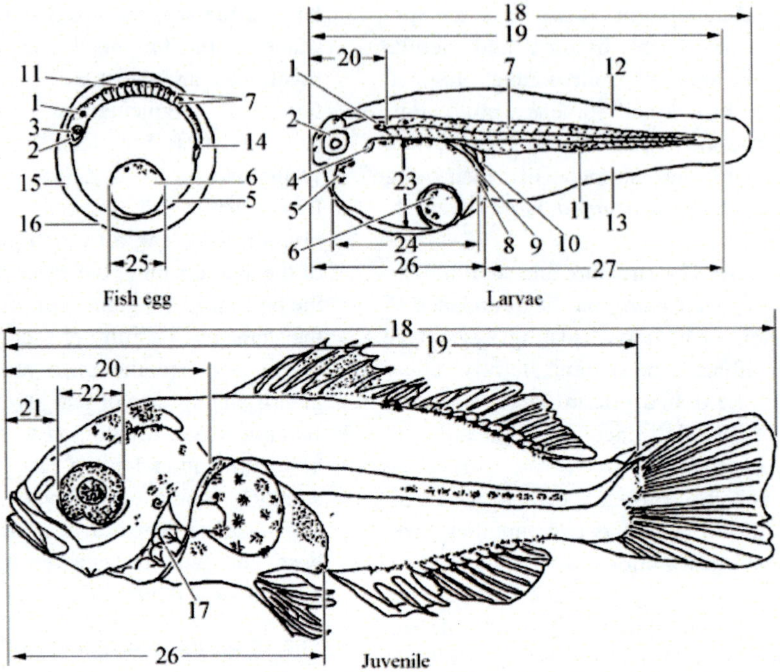

Fig. 3.4 Site and name of determination points for eggs, juveniles (Zhang et al. 1985)
1 auditory sac, *2* eye, *3* crystal, 4 Heart, *5* yolk (egg), yolk sac (litter), *6* oil bulb, *7* myotome, *8* anus, *9* digestive tract, *10* bladder, *11* melanopsin, *12* dorsal fin membrane (dorsal fin fold), *13* ventral fin membrane (ventral fin fold), *14* Gram's vesicle, *15* perivitelline space, *16* egg membrane, *17* anterior gill cover bone outer margin spine, *18* full length *19* body length, *20* head length, *21* muzzle length, *22* eye diameter, *23* yolk sac short diameter, *24* yolk sac long diameter, *25* olecranon diameter, *26* anterior anal distance, and *27* posterior anal distance

Embryos in live eggs, live young, and juvenile fish often exhibit bright colors, but soon after death, the colors fade; melanocytes are all that remain, and the time of appearance, number, size, shape, and location of melanocytes are the main basis for identification of the species. When live specimens are observed, all kinds of pigment cells are described, while for fixed specimens, melanocytes are mainly described.

3.4 Analysis of Environmental Factors Affecting the Survival of Juvenile Fish

Changes in the distribution and abundance of marine fish spawners and eggs in the pelagic zone and their correlation with changes in environmental factors are one of the main bases for predicting replenishment and its variability. Quantitative environmental factors, mainly water temperature, salinity, depth, current speed, wind, and waves, as well as water pollution and pH (pondus hydrogenii), have direct and indirect effects on the distribution and survival of fish eggs and juvenile fishes. Eggs and juvenile fishes are the youngest and most vulnerable stages of the fish life history process, and any unsuitable environmental conditions can cause significant mortality. For example, the development and growth of various species of eggs and juvenile fishes require a suitable temperature range, and unsuitable changes in water temperature will retard their development and even lead to mortality; their distribution is also necessarily limited by isotherms. If currents carry them to waters that are not suitable for development, then this will lead to their death. This idea has gained wide acceptance as an important factor in the mortality during egg and smolt stages and is also supported by many studies. Changes in environmental physicochemical factors are usually considered to have the most dramatic and pronounced effects on the number of offspring and the number of replenished populations of spawning fish in estuaries. For example, the Argentine shortfin squid *Illex argentinus*, which is distributed in the southwestern Atlantic, lives for 1 year, dies after spawning, and has no remaining population, and its resource recruitment varies drastically yearly. It was found that its stock recruitment is very closely related to the marine environment of the spawning grounds during the previous year. The higher the range of water temperatures suitable for spawning and its specific gravity during the spawning period, the higher the stock recruitment is the next year. This scenario has been confirmed during actual production.

Environmental physicochemical factors are of particular importance because they influence the distribution and density of prey organisms and thus the distribution and survival of juvenile fishes. It has been found that the survival of juvenile fishes is dependent on the presence of small dense areas of prey, or "patches." Once a school of juvenile fishes finds a prey "patch," it has the ability to stay in the "patch" to feed. The distribution of young fish and their prey organisms in the ocean is not random but rather unevenly distributed in dense areas. The results of many studies support this notion.

New methods and tools for investigating the distribution of eggs and juvenile fishes in the ocean, their population changes, and their correlation with environmental factors are constantly being developed, and it is now possible to use tools such as ocean satellites to infer the concentration areas of eggs and juvenile fishes in the ocean and to sample quantitatively at different water depths. Therefore, the study and determination of the effects of various environmental factors (e.g., water temperature, salinity, flow rate, and pH value) on juvenile development, survival in natural habitats, etc., in conjunction with actual field surveys, are of great significance and practical guidance for exploring the causes of early fish mortality, protecting natural resources and developing indoor factory nurseries.

References

Chen DG (1997) Biology of fishery resources. China Agricultural Press, Beijing. (In Chinese)

Chen XJ (2014) Fisheries resources and fishery oceanography. Ocean Press, Beijing

Chen XJ, Liu BL (2017) Biology of fishery resources. Science Press, Beijing. (In Chinese)

Yin MC (1995) Fish ecology. China Agricultural Press, Beijing. (In Chinese)

Zhang RZ, Lu SF, Zhao CY et al (1985) The eggs and young of fishes offshore Chinese. Shanghai Science and Technology Press, Shanghai. (In Chinese)

Age and Growth of Fish

Xinjun Chen, Bilin Liu, and Zhou Fang

Abstract

Age and growth are the basic contents of fishery biology research. Because of the seasonal changes in the aquatic environment, including the biological and nonbiological environment, there will be obvious marks in the hard tissue materials of the fish body, such as scales, vertebrae, otoliths, and other hard tissues. These marks can be used as the discrimination basis of the annulus and the daily increment; at the same time, according to the marks left on the hard tissues of fish body, we can study their life history, growth speed, age of sexual maturity, spawning period, and spawning habits, so as to provide basic materials for the study of fish fishery biology. In this chapter, the significance of the study of fish age and growth is briefly described, and the general principles of annuli formation and general materials for aging are described in detail. The structure, types, circulus characteristics of scales, and the matters needing attention in collecting scales are also introduced. At the same time, other identification materials, such as fish otolith, and aging methods, including crustacean aging and cephalopod aging methods, are also described in detail. Moreover, other methods for studying other age of fish, such as mark-release method, fish age determination based on body length frequency, isotope verification method, etc., are also presented. In this chapter, we also systematically described the growth of fish and its determination method, and introduced the study of age and growth based on statolith microstructure of *Illex argentinus* in the southwest Atlantic Ocean. The emphasis of this chapter is to master the methods of fish age identification and growth calculation.

Keywords

Age · Growth · Age determination

Abbreviations

AIC	Kaike's information criterion
Ap	Apical region
AGR	Bsolute growth rate
a	Biologically significant and the body length at the time of scale appearance;
α	The rate of assimilation
β	The rate of substance decomposition per unit weight; the rate of anabolism
b	Corresponds to the body length per unit of scale.
Ba	Basal region

X. Chen (✉) · B. Liu · Z. Fang
College of Marine Sciences, Shanghai Ocean University, Lingang Newcity, Shanghai, China
e-mail: xjchen@shou.edu.cn; bl-liu@shou.edu.cn; Zfang@shou.edu.cn

X. Chen, B. Liu (eds.), *Biology of Fishery Resources*, https://doi.org/10.1007/978-981-16-6948-4_4

BW	Body weight
CCD	Charge-coupled Device
Ci	Circuli
CNS	Central nervous system
CL	Crest length
FAPs	Fluorescent age pigments
G	The relative growth rate
HL	Hood length
H	The rate of substance synthesis per unit of "physiological surface";
K	The relative marginal growth value; growth parameter related to fish metabolism and growth; Laateral region
LWL	Lateral wall length
L_t	The length of the fish in previous years; individual length at t;
L_0	The measured body length at the time of capture
L_1	The length and weight of the fish at the beginning of the period for which the growth ratio is calculated
L_2	The length and weight of the fish at the end of the period for which the growth ratio is calculated
L_∞	Asymptotic length
L (p_1,..., p_m)	The maximum likelihood value of the mantle length at the daily age
ML	Mantle length
r_t	The length of the scales in the year corresponding to L_t;
r_n	The length of each circulus
$r_n - r_{n-1}$	The distance between the penultimate first and second rounds
R_2	The body weightor mantle length at age t_2
R_1	The body weight or mantle length at age t_1
R	The scale length
R-r_n	The distance from the edge to the penultimate first round;
RL	Nrostrum length
RSS	Rostrum sagittal section
SF	Scale focus
Sg	Scale groove
S	The effective physiological surface of the organism;
t_0	Hypothetical constant; theoretical age at $L = 0$
t	Age in day
t_1	The time at the beginning of the period for which the growth ratio is to be calculated
t_2	The time at the end of the period for which the growth ratio is to be calculated
WL	Wing length
W	Body weight
W_t	Individual weight at t
W_∞	Asymptotic weight
σ^2	The variance of the error term

4.1 Significance of Studying Fish Age and Growth

Age and growth are fundamental components of fishery biology research. Due to seasonal changes in the water environment, both biotic and abiotic, distinctive markings are left in the hard tissues of fish (e.g., scales, vertebrae, otoliths, fin rays, cleithrum, and operculum), which can be used as a basis for discriminating annual and daily rings and for studying the past life of a fish and its growth rate, age of sexual maturity, and spawning date and spawning habits, thus providing a basis for research on fishery biology.

Fish age and growth studies are fundamental to the assessment, scientific management, and sustainable use of fishery resources. Studying fish age and growth has the following main implications:

(1) These studies can provide a scientific basis for setting reasonable fishing intensity levels. The purpose of fishery production is to be able to obtain a certain amount of quality catch from waters in a stable manner over a long period of time and to ensure the sustainable use of the fish. The basic indicators for determining an optimal catch are as follows: first, it must be a large and good quality catch, and second, the appropriate growth rate of the fish must be caught so that it meets fishing commodity specifications. Optimal catches

are generally considered to occur in pristine waters where there are slightly more older fish and a certain proportion of fish in each age group and in underutilized waters. Conversely, waters that have been fully exploited or are especially overfished have underage age groups, and these waters also have fish with a delay in the date of the first sexual maturity stage and a small proportion of older fish.

There is a relationship between the growth rate of fish and the amount of resources contained in their habitats. If there is no change in prey in the waters and the number of fish increases, then this scenario will inevitably affect the growth rate of fish, delaying sexual maturity and resulting in smaller body lengths, which are not conducive to increased catches. In contrast, an appropriate number of fish is conducive to reasonable foraging for prey, rapid growth, and increased weight at sexual maturity, which are conducive to an increase in catch. That is, fish resource reserves are inversely proportional to the length of the fish body.

(2) These studies can also provide a scientific basis for determining reasonable fishing specifications. Limiting catch sizes in fishing waters is very important and is one of the key elements of fisheries management. The size of a fish at the first sexual maturity and the first entry into the fishing population depend on the growth rate of the fish. It is generally accepted that an excessive number of older fish are not conducive to the proper use of prey in waters because older fish grow slowly and are not conducive to increasing the productivity of waters. The best fish growth rate is achieved during the period of the fastest growth rate, when the consumption of prey in the waters is reasonable, which is a principle promoted by the aquaculture industry and is also a principle of fisheries resource management.

(3) These age and growth studies can provide basic information for fishing forecasting. By accumulating long time series of fish catches and age composition and by mastering the relationship among growth patterns, fisheries, and the environment, we can obtain an understanding of the biological characteristics of fish so that we can scientifically prepare catch forecasts based on fish age composition and growth, combined with environmental factors, and provide the basis for fishery forecasts.

(4) These studies can provide scientific data and measures for fish farming. Based on the growth characteristics of fish, especially the demand for prey, growth rate and needed environmental conditions, the species and quantity of fish that should be cultured in water, as well as the interaction between species and feed supply can be determined to improve the quality and yield of cultures.

(5) These studies can provide a basis for improving the effectiveness of species translocation and domestication. The growth characteristics of fish allow the identification of growth rates, patterns, and prey requirements of fish and thus the needed improvements in environmental conditions to accommodate the growth, development, and reproduction of fish to increase the number of new species for translocation and domestication, enhance the domestication effect, and increase their commercial value.

(6) These studies can provide a basis for fish stock identification. The growth characteristics of fish are also important factors for identifying fish populations. For example, the wide distribution of pollock in the northwestern Pacific Ocean, with differences in age composition, morphological characteristics, and ecological habits, has enabled the identification of three local populations of pollock, namely, the Bering Sea, Okhotsk Sea, and North Japan Sea populations. The North Japan Sea populations spawn offshore in spring; this main population consists of 5- to 6-year-olds and is larger than the other two northern populations.

4.2 General Principles of Annulus Formation and Its Determination Materials

4.2.1 General Principles of Annulus Formation

Fish growth is similar to the growth of most vertebrates in that two changes occur during their growth: the body grows and body mass develops. Usually, these two growth phenomena occur simultaneously and are complementary. Body size growth is the process of increasing body length and weight, while body mass growth involves the process of gonadogenesis and sexual maturation. The growth of a fish does not end when it reaches sexual maturity; a fish continues to reproduce until it dies of senescence and terminates. There are exceptions, such as salmon, which grows only in shape and shrinks in size during spawning migration, so the growth of this fish is uneven. The life span of salmon is 2–4 years, and after spawning, the parent fish dies of physical exhaustion. The growth of most fish is balanced, i.e., a fish continues to grow over time in terms of body mass. This characteristic of fish growth is mainly determined by nutritional conditions.

When fish consume many nutrients in summer, their growth is very rapid, while in winter, when fish lack food, their growth rate slows or even stagnates. This fish growth pattern is reflected in the growth of scales, bones, and other hard tissues; that is, fish grow very rapidly in spring and summer, forming many concentric circles on scales and other hard tissues, in a loose condition called a "sparse belt," also known as a "summer ring," while in autumn and winter, fish grow slowly or even stagnate. During this time, the concentric circles formed on the scales and other hard tissues are narrower and called "dense bands" or "winter rings." The sparse and dense rings combine to form a growth ring so that each year, a growth ring is formed, which is an age ring or an annual ring. Usually, most fish follow the above rule, but there are some exceptions; for example, in the East China Sea, *Dentex tumifrons* forms two annuli a year; when yellow bream grows to an advanced age, its annulus formation is irregular, and sometimes only one ring appears in 2–3 years. Therefore, the formation of fish annuli is seen as a result of not only seasonal water temperature changes but also changes in the process of fish growth (such as the role of genetics) through internal physiological mechanisms due to cyclical changes in the external environment, which is the result of physiological cyclical changes in fish.

4.2.2 Materials for Determining the Age of Fish

Usually, the materials used to determine the age of fish are scales, otoliths, operculum, vertebrae, fin rays, cleithrum, etc., and the tools used for observation are microscopes, dissecting microscopes, etc. Since the habits and tissues of fishes are different, the best materials for fish age determination are also different; therefore, several materials are usually used for identification and comparative analysis before determining their ages. For example, the following materials are used to identify the age of some Chinese economic fishes: otoliths are used as the main material, and scales are used as a supplement for *Larimichthys crocea* and *Larimichthys polyactis*; otoliths are used as the main material, and vertebrae are used as a supplement for *Trichiurus lepturus*; otoliths are used as the main material, and vertebrae and scales are used as a supplement for *Scomber japonicus*; scales are used as the main material, and otoliths are used as a supplement for *Ilisha elongata*; scales are used as the main material for *Decapterus maruadsi*; scales are used as the main material for *Sardinops sagax*; and scales are used as the main material, and otoliths are used as a supplement for *Clupea pallasi*.

Age determination of fish is a sensitive and fundamental task, especially for older fish, and is often quite difficult; thus, age determination of fish usually requires two or more people to read independently and then to compare the results. Errors of more than 10% need to be reread;

otherwise, the average can be taken. Age determination of fish and invertebrates can be conducted by rearing, marker release, observation of annuli, and analysis of length distribution. The more applicable methods are the annulus method and the length frequency method. The chronology method in turn includes the use of hard tissues such as scales, otoliths, vertebrae, etc. for age determination.

4.2.3 Expression of Age Groups and Basic Concepts

Studying fish age is an essential activity of research on the fundamentals of fish biology and aquatic resource surveys, and therefore, it is extremely important to accurately identify and delineate the age groups of fish. Some common fish age group concepts are described as follows:

(1) Yearling fish. Current-year fish are small and fully formed with scales (usually from the second half or autumn of the year in which the fish's life begins), and no signs of annuli appear on the scales. Fish in this group are represented by age group zero (0).
(2) Winter-aged fish. Winter-aged fish are current-year fish that have overwintered and have completed the first stage of growth. The name "winter-aged fish" can also be used in the spring for fish hatched the previous fall, and winter-aged fish may be less than a full year old, usually with an annual mark on their scales. This group of fish is also called the first age group (I).
(3) Second summer-aged fish. A second summer-aged fish is one that has lived over two summers, and this name is used from the second half of the second year and the autumn after the beginning of the fish's life. The scales bear a trace of an annulus with a generally partial ring of the second year's increase on the periphery of the ring. Second summer-aged fish also fall into the category of the first age group (I).
(4) Second winter-aged fish. The second winter-aged fish is a second summer-aged fish that has overwintered and has 2 years of scales, with either one annulus or an almost completed second year of accretion. However, the second annulus does not yet appear on the edge of the accretion ring. Sometimes there are several broad, bright rings of the third-year verticillation at the periphery of the second verticillation. Depending on the width and sparseness of the rings and the appearance of the entire growth band (narrow ring), this new growth ring is easily distinguishable from the ring of the previous year, which is fully grown.

In the spring or first half of the third year, the scales have two annuli and a few rings of the accretionary part of the third year. Second winter-aged fish belong to the second age group (II).

(5) Age determination of fish. The age of a fish is the number of years it takes to complete its life cycle or the number of years it has lived. Individuals of the same age in a school of fish are called cohort fish. In statistics, these fish of the same age are grouped together and referred to as the same age group; e.g., fish born in the current year are referred to as age group 0. Fish in their second year of life are called age group 1, fish in their third year of life are called age group 2, etc. A group of fish with all individuals born in the same year or season is referred to as occurring in the same generation. The year of birth is generally used to determine the particular generation of a fish. If that generation occurs in extremely sufficient numbers, that is, the parents are abundant, then spawning is high, the juveniles are in a good environment during their developmental stage, prey is abundant, and survival rates are high, thus constituting a rich catchable resource. A high number of fish in a catchable generation are called strong generation. For example, in 1971, *Engraulis ringens* belonged to the strong generation category, resulting in a high level of anchovy production that was more than 12 million tons in 1972. The ratio between the number of

individuals in each age group and the number of all individuals is known as the age composition of the catch. Some fish have a large number of age groups, which can be more than 20, such as *L. crocea*; other fish have very few age groups, such as the jack mackerel; and others have only 1–2 age groups, such as sardines and anchovies.

(6) Identification of chronological records. In general, the actual age of a fish is rarely a whole number, but in studying the age of fish populations, it is not necessary to know the precise number; thus, it is customary to count fish by terms such as "n-age fish" or "n-age group." To indicate new growth outside a ring after the formation of the annulus, a "+" sign is often added to the upper right corner of the annulus number, such as 1^+, 2^+, 3^+, ...n^+.

4.2.4 Comparison of Fish Age Determination Using Scale and Otolith

In China, scales have long been regarded as reliable aging materials, especially in freshwater fishes, except for a few species with no scales or no obvious annulus features on their scales; usually, only scales are used as aging materials, but occasionally, some vertebrae and fin rays are used as supporting evidence for aging. However, some researchers have pointed out that determination by scales can be confusing and that annuli can sometimes be overlooked or mistaken for checks, resulting in errors in age identification. Many studies have shown that scales are only suitable for determining the age of faster-growing, younger fish and that in comparison to otoliths, fin rays, vertebrae, cleithrum, operculum, and scales usually underestimate the age of older and slower-growing individuals with less accuracy and precision, with sometimes large gaps. Otoliths are considered to be a more reliable and accurate material than other calcified structures.

Therefore, it is generally considered feasible to use otoliths and scales for populations with a simpler age composition and faster growth, and scales have the advantage of being easy to collect and simple to handle. In cases where accuracy is not very important, only scales can be taken as material for age identification, whereas for populations with complex age compositions and slow growth, the choice of scales as an identification material is clearly not applicable, and otoliths are used instead.

4.3 Scale Structure and Fish Age Determination

A scale is a derivative of fish skin, a construct adapted to the aquatic environment. It is widely found on the body surface of extant bony fishes as the exoskeleton of fishes. The number and morphological characteristics of scales are one of the main features of fish classification and an important basis for age identification and growth status analysis. Since the 1970s, scanning electron microscopy has been applied to observe the structural features scale surfaces, and this technique is of great importance to the systematic study of fish taxa and population characteristics. Instruments used to observe scales are usually microscopes, dissecting scopes, projectors, photographic magnifiers, and slide projectors.

4.3.1 Structure of Fish Scales

The scales of fish are mainly divided into three parts: lateral line upper scales, lateral line lower scales, and lateral line scales, whose structure directly reflects the taxonomic characteristics and growth characteristics of fish and is an important part of studying fish classification, survival environment, and growth trends. The part of each scale toward the head of the fish and buried in the scale capsule is the anterior zone of the scale, the part toward the tail of the fish and exposed outside the capsule is the posterior zone, and the part located between the anterior and posterior zones is the lateral zone. The growth of the annulus is closely related to the growth rate of the fish body, and scales are composed of broad and narrow bands arranged one after another as they grow (Figs. 4.1, 4.2, and 4.3). Scale structure can be described as follows:

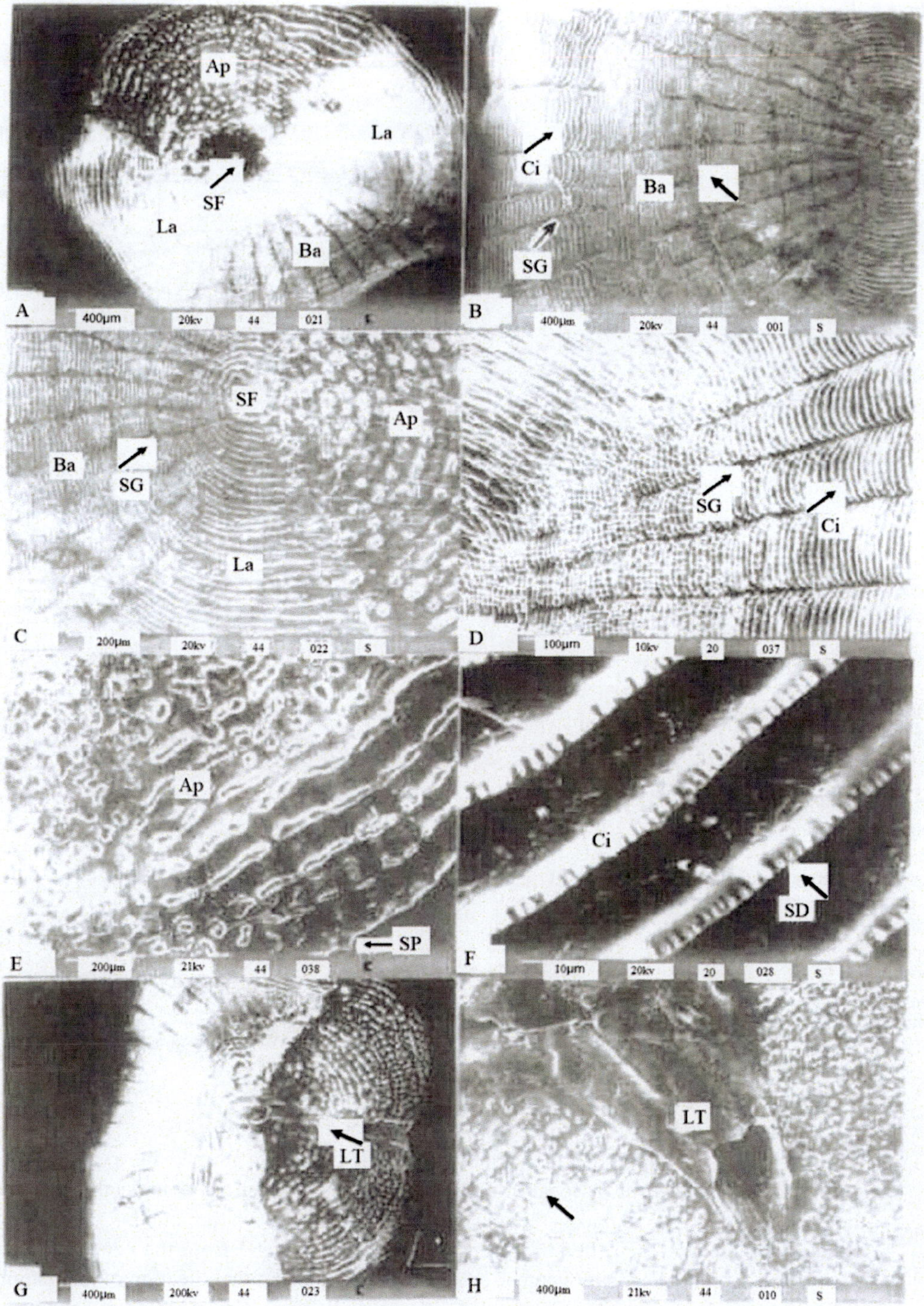

Fig. 4.1 Schematic diagram of *Oreochromis aureus* scales (Fan et al. 2000)
A: Structure of *O. aureus* scales with Ap as the apical region, La as the lateral region, Ba as the basal region, and SF as the scale focus
B: Structure of the basal region of the scale, SG for the scale groove, Ci for the circuli, (→) for the annulus
C: Structure of the scale focus
D: Magnified images of the basal region
E: Magnified images of the parietal region, SP for scale spines
F: High magnification images of the circuli

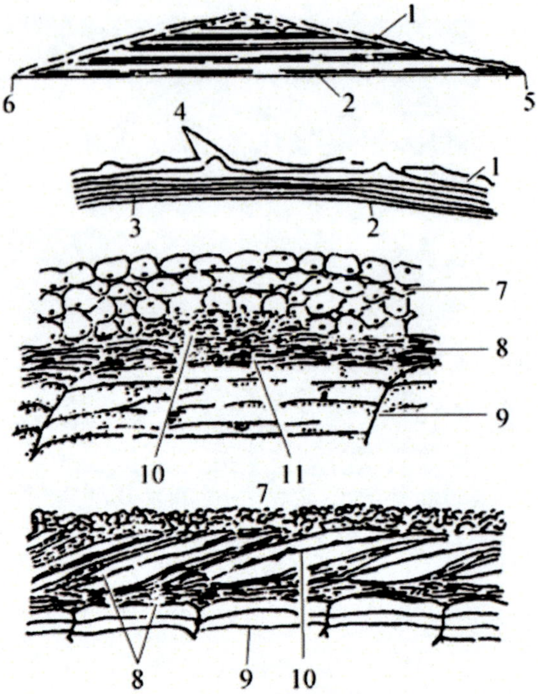

Fig. 4.2 Cross-section of scales and early formation of male scales (Chen 2014). a. Circulus, b. basal, c. underside, d. dorsal, e. posterior region E. epidermal cells, D. true epidermis, M. subcutaneous layer S. scales, SP. scutellum

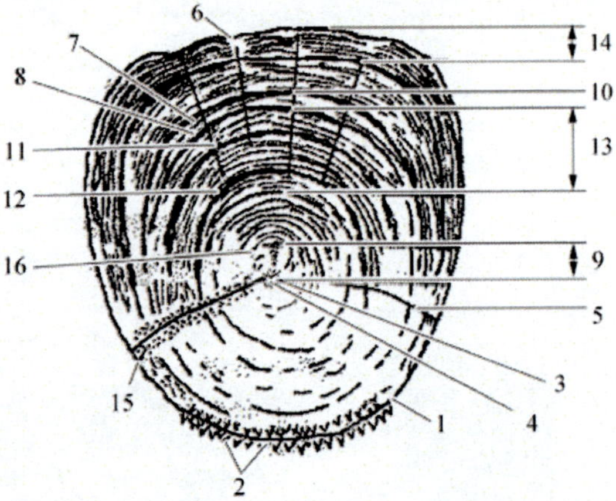

Fig. 4.3 Pattern diagram of fish scales (Chen 2014). a. Posterior region, b. ctenidia, c. focus, d. central region, e. lateral region, f. anterior region, g. circulus, h. annulus, i. space of young annulus, j. reproductive annulus, k. additional annulus, l. first annulus, m. age spacing, n. marginal spacing, o. verrucae, and p. young annulus

(1) Scale groove. This is a depression in the bony layer of a scale. It is formed by local folding of the bone, and this groove makes the scales bend easily, increases their flexibility and elasticity, and is suitable for movement and nutrient transport. In the cuticle of the scales, this groove is formed by a special shallow fracture, which is completely revealed by the presence of the annulus. Usually, radial grooves extend from the center of scales or slightly off in the direction of the margin in a radial arrangement. Radial grooves are

Fig. 4.1 (continued) **G**: The structure of the scale of lateral line, LT for tube of lateral line **H**: The high magnification image of the tube of lateral line; (→) Body of scabrosity different in different species; for example, salmon scales have no radial grooves; cod scales have radial grooves arranged by small pillow-like circuli, with their respective separated interstitial grooves distributed throughout the scales; and herring scales and snapper scales have well-developed radial grooves. Some fish have radial grooves that radiate in all directions, e.g., Cobitidae and Periophthalmidae; have radial grooves that radiate only in the anterior region, e.g., Sparidae and *Gambusia affinis*; have radial grooves that zigzag, e.g., *Tenualosa reevesii*; and have radial grooves that form a circle, e.g., *Misgurnus anguillicaudatus*.

(2) Circuli. Circumferential lamellae are the raised lines of the bony layers on the surfaces of scales. The characteristic arrangement of the annulus on the scales corresponds to seasonal changes in the habitat of a fish and changes in the physiological condition of a fish (e.g., sexual maturity status and blood calcium content), and this arrangement reflects the growth of the fish in past years. Many elevated lines are arranged around the center of the scale focus, and these elevated lines are called circuli or increments. The arrangement of the increments is generally in the form of concentric circles, but there are also rectangular increments or other shaped increments, which differ mainly according to scale type. The structure of the increments on fish scales is further divided into annual, juvenile, additional, and reproductive scars.

(3) Scale focus. The scale focus is in the center of the scale and is the earliest part of the scale formation. The surface structure of scales expands in all directions with this as the center, and annuli are usually formed on two sides first and then gradually surround the front area to form a complete annulus. The nucleus is in the center of the scale or on one side; for example, the nucleus of *L. crocea*, *L. polyactis*, *I. elongate*, and *Hypophthalmichthys molitrix* is located in the center of the scale, while the nucleus of *Parabramis pekinensis* is located in the posterior area of the scale. The position of the focal nucleus depends on three factors: first, whether the scales grow on the same growth axis or the growth axis is somewhat off; second, the original condition of the growth center is different and can determine scale size and the position of the scale focal nucleus; and third, the sizes of the scales differ when they are buried in the anterior zone.

4.3.2 Types of Fish Scales

The study of most bony fish scales has led to the classification of scales into four representative types (Fig. 4.4).

(1) Salmon trout type. The annulus is arranged in concentric circles with the scale tarsi as the center. Depending on the fish, the locations of the scales vary, and the shape of the scales also varies slightly. The scales are thin and without radial grooves. The annular pattern is sparse and dense, with remarkable regularity. Fish that have scales of this type include *Oncorhynchus mykiss*, *Salmo salar*, etc.
(2) Snapper type. Scales are rectangular and slightly right-angled from left to right at the anterior end, with many notches on the edges of the anterior area. The annulus is arranged with the scale tarsi as the center, forming many similar rectangular circles. There are distinct "transparent rings" between the rings. The distance between the annuli gradually decreases toward the outer ring. From the scale tarsi toward the anterior margin, a radial spoke groove forms. Fish that have main scales of this type are *L. crocea*, *L. polyactis*, *Pagrus major*, *Dentex tumifrons*, *Acanthopagrus latus*, etc.
(3) Herring type. The scales are rounded, thin and translucent, densely covered with microscopic rings, and sparsely arranged to intersect the median axis at almost right angles. The radial grooves diverge from the central radius to the sides, similar to birch branches. The annuli are very clear and appear in concentric rings. Fish with scales of this type

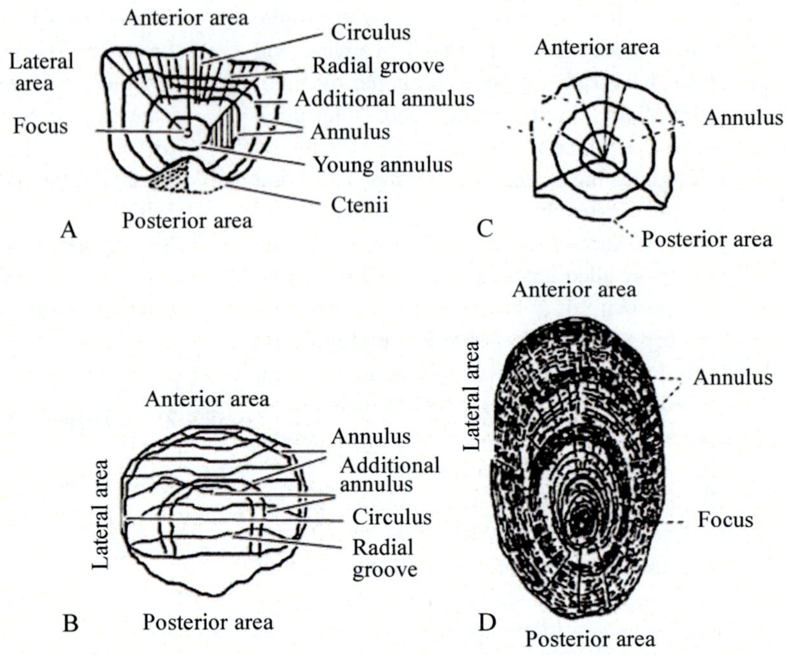

Fig. 4.4 Fish scale types (Chen 2014). **(A)** Snapper type (*L. polyactis*); **(B)** herring type (*C. pallasii*); **(C)** salmon trout type (*Oncorhynchus keta*); and **(D)** cod type (*T. chalcogramma*)

include *I. elongate*, *Clupea pallasii*, *S. sagax*, *Coilia nasus*, and *C. mystus*.

(4) Cod type. The scales are small and elliptical, and the annulus is also arranged on the scales in a concentric shape, which is composed of many small pillow-like protrusions. The rings of the annual ring are arranged in a sparse pattern of annuli, especially in the posterior area of the scales, which is clearer. Fish with scales of this type include *Gadus macrocephalus*, *Theragra chalcogramma*, *G. morhua* , etc.

4.3.3 Characteristics of Fish Scales Annuli

Generally, the scales annuli of bony fishes can be grouped into the following five morphological character types (Fig. 4.5):

(1) Sparse and Dense Type

The circulus forms wide, sparse growth bands as well as narrow, dense growth bands, and the junction of the narrow and wide bands is the annulus. During the spring and summer, the fish have a very vigorous metabolism and grow rapidly, forming wide rings on the scales, while during the winter, growth slows down and forms dense rings, alternating between the two. Then, growth is repeated the following year, leaving the wide, dense rings on the scales for the second year. In addition, this process continues for several to a dozen years. In approximately the last decade of a fish's life, we can still find these annuli on the scales, but the distance between the rings becomes increasingly shorter until it is difficult to distinguish between them.

This sparse and dense arrangement of scales is characteristic of the vast majority of fish, such as *L. crocea*, *L. polyactis*, *Nibea albiflora*, *P. major*, *C. nasus*, and *Paralichthys olivaceus*.

(2) Cutting Type

When growth is normal, the circulus is arranged in concentric circles. When growth is slow, the circulus is not round but gradually shortened, and its two ends terminate in different parts of the posterior lateral area of the scale. When growth resumes the next year, a new circulus grows along the whole edge of the scale again, forming a complete circulus and the different directions of the circulus group; that is, the

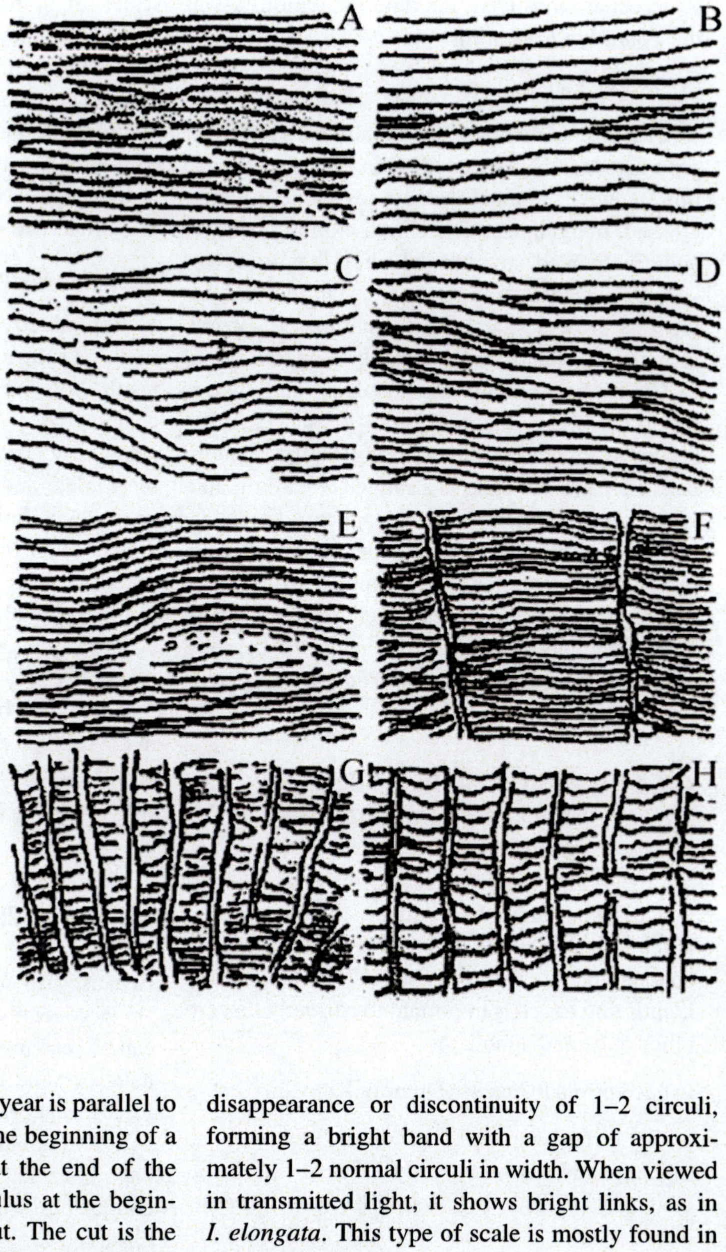

Fig. 4.5 Morphology of annuli on the scales of several fish species (Chen 2014). **(A)** *C. pallasii*; **(B)** *O. keta*; **(C)** *S. japonicus*; **(D)** Beryciformes; **(E)** *P. major*; **(F)** *Branchiostegus japonicus*; **(G)** *Jaydia lineata*; and **(H)** *Sphyraena pinguis*

arrangement of the circulus in 1 year is parallel to that in the following years. At the beginning of a new year, the circulus group at the end of the previous year and the first circulus at the beginning of the new year form a cut. The cut is the annulus, and it is generally clearest at the junction of the top and side areas of the scales. Fish with this type of scale include *Saurida elongata*, *H. molitrix*, and *Cyprinus carpio*.

(3) Bright Type

Due to incomplete development of the circulus on the annuli of scales, there is often a disappearance or discontinuity of 1–2 circuli, forming a bright band with a gap of approximately 1–2 normal circuli in width. When viewed in transmitted light, it shows bright links, as in *I. elongata*. This type of scale is mostly found in the anterior region.

(4) Flat and Straight Type

As the arrangement of circuli on scales is generally curved under normal growth, suddenly 1–2 circuli appear in a flat arrangement, distinct from the adjacent circuli; i.e., two annual circuli are separated by a flat arrangement. This type of scale occurs mostly in the anterior region, as in the *Pennahia argentata*.

(5) Messy Pattern Type

The arrangement of the circuli between two growth zones is disorganized in terms of direction, uneven, and sometimes intermittent, crossed, merged, etc. The annuli exhibit a sparse and fragmented structure of circuli, with occasional sparseness and cutting. Near the end of the first growth year band (sometimes a second growth year band also appears), there are often 2–3 circuli close to each other, and when observed under magnification, they generally appear as thick black linear shadows; the remainder of the growth band is arranged as circuli, and wherever a broken structure appears, it is an annulus. Some of the circuli appear intermittently in a wavy pattern, and some appear to cross or merge into punctate lines. Such features occur more frequently in the anterior or lateral regions of scales, such as in *Squaliobarbus curriculus*.

4.3.4 Additional, Reproductive Checks and Regenerating Scales of Fish

In addition to annuli, other rings, such as addition, juvenile, and reproductive checks, are also present on the scales of fish. Therefore, when identifying age, it is important to correctly discern other rings and annuli.

(1) Additional Checks (Dummy Rings)

False rings are marks left on the scales of fish that are suddenly and greatly affected by insufficient prey, changes in water temperature, disease, etc., during their normal life. In general, the additional checks are not as clear as the annuli and are fragmented rings. It needs to be verified by annual observation or by means such as back-calculation in relation to scale length and body length.

(2) Juvenile Rings

Juvenile rings are also one of the additional checks, a small ring in the central area of the scales of some fish, also known as the "zero ring," and they are most easily confused with the first annuli. The juvenile ring can be judged according to the back-calculation method of the relationship between scales and body length, combined with the analysis of biological characteristics of fish such as anadromy, depth of habitat, and change in food habits.

(3) Reproductive Checks

Reproductive checks, also known as spawning rings or spawning marks, are rings formed as a result of reproductive action. They are characterized by broken, divergent, and irregularly arranged annuli in the lateral areas of the scales, and the top area of the scales often produces a darkened annulus that has thickened and is often broken or has many small, curved sections, with the sides of the annulus often immediately above a structureless, shiny gap.

(4) Regenerative Scales

Scales can be shed, and new scales grow back in the original area. The central part of such scales is no longer visible as a regular ring and is not suitable for age determination.

4.3.5 Collection of Fish Scales

Because fish scales are relatively easy to obtain, are large relative to materials such as otoliths, and can be seen more clearly for aging samples, scales have become the most commonly used material for fish age identification. In general, before the nature of annulus formation of a particular fish is known, zonal collection of scales should be conducted on the side of the fish body and then observed for comparison, selecting the area with regular scales and distinct rings as the site of scale collection. After the scales are removed, they can flake immediately. To obtain a slice, fresh scales can be dipped in light ammonia or warm water for a few minutes, and then, the surface can be gently rubbed with a toothbrush or soft cloth, put into clean water and rinsed, wiped dry, and observed.

4.4 Fish Otoliths and Age Determination

4.4.1 Discovery of the Otolith Daily Increment of Fish, Progress in Its Study, and Its Significance

In 1971, Panella with the Department of Geology and Geophysics at Yale University first proposed the existence of daily increment in otoliths in *Merluccius bilinearis*, and some scholars successively confirmed the existence of daily increment in otoliths in other fish species, such as Clupeiformes, Salmoniformes, Myctophiformes, Anguilliformes, Cypriniformes, Cyprinodontiformes, Gadiformes, Perciformes, Pleuronectiformes, and more than a hundred species of marine and freshwater fishes. Thus, otolithic daily increments are a common phenomenon in fish. In addition, the linear relationship between the width (spacing) of the otolithic daily increment and body weight has been used to project the body weight and growth of juvenile *Oncorhynchus nerka*, and the ratio of strontium and calcium content in otolithic daily increments has been measured using an electronic microprobe as an indicator of environmental history changes to study the life history of fish. The discovery of the otolith daily increment is the most important progress in fish biology research since the 1970s, and it has broadened and deepened the field of fish biology research. As a result, fish otoliths are commonly used to identify the age or ecological taxa of fishes (especially marine fishes).

The study of otolith daily increments has a broad development prospect, especially in terms of the isotopic analysis of otoliths and their chemical composition and microstructure, and the early life history of otolith daily increments and fish may become popular topics. However, the study of otolith increments is an emerging field, and there are still many questions related to the formation process of subdaily increments and transitional and polycentric daily increments; the rate of daily deposition; the main factors affecting the formation of daily increments; and the mechanism of daily-increment formation, all of which need to be studied in depth.

The otolith daily increments reveal the relationship between the growth and development of fish and the external environment, which has not only theoretical significance but also important application value. First, otolith daily increments can accurately describe the growth of fish, and using daily age as the time unit to describe the growth of fish can objectively reflect the growth characteristics of fish. Second, otolith daily increments can provide information on the life history of fish, and otolith daily rings have certain environmental sensitivity, so they can be used to trace changes in fish habitat according to the change in daily-increment spacing. Third, otolith daily increments can facilitate the study of fish population ecology and fishery resources, and more accurate and reliable results can be obtained by using otolith daily increments to study the replenishment rate and mortality rate of a population and to identify different breeding groups.

4.4.2 Morphological Characteristics of Daily Increments in Fish Otolith

In the inner ear of fish, the utricular vestibule, the saccular vestibule, and the lagenar vestibule have a pair of lapillus, sagitta, and asteriscus, respectively, on which the daily increment is deposited. Sagittal otoliths are larger than the other otolith types in most fish, so sagittal otoliths are generally used as material to study daily increments. However, some scholars believe that the lapillus otolith of Cyprinidae has more stable morphological variation and is more suitable for daily-increment growth studies, as shown in Fig. 4.6, which presents a schematic diagram of the determination of increment width in the otoliths of *Mugil so-iuy* paralarvae.

When production is observed under light or scanning electron microscopy, the otolith is centered on a nucleus with concentrically arranged daily increments outside the nucleus. As the otolith morphology changes with the growth of a fish, the daily-increment morphology changes

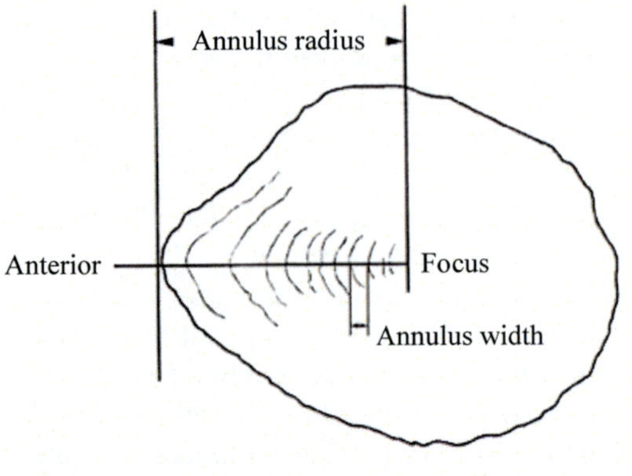

Fig. 4.6 Schematic diagram of the determination of otolith increment width of *Mugil so-iuy* paralarvae (Li et al. 1993)

accordingly, generally from concentric round rings in initially orthoclase otoliths to concentric pear-shaped or oblong increments in eventually stable pear-shaped or oblong otoliths. When an otolith changes from a square circle to a pear or oblong shape with one end round and one end slightly pointed, its center is located near the end of the circle, and the otolith has a long and short radius. Usually, the short-radius daily increments are closely arranged and clear, while the long-radius daily increments are more sparsely arranged and have more disordered and unclear segments of increment strands; thus, the daily increments are mostly measured by short-radius counting. Under a transmission light microscope, an increment is composed of a transparent growth band and a dark intermittent band. The ultrastructure shows that the growth band consists of aggregates of needlelike calcium carbonate crystals, the interstitial band is organic filler, and the two bands interpenetrate each other. In some fish otoliths, in addition to the normal daily increment, there are thicker and more distinctly marked increments due to changes in the developmental stage or ecological conditions of the fish, and in some fish yolk-nourished or mixed-nourished juveniles, daily increments appear as slender subdaily increments.

4.4.3 Daily-Increment Growth Pattern of Fish Otoliths

Otolith daily-increment studies first confirm whether the rings on the otoliths form one ring a day. This information can be obtained by using the daily-age versus daily-increment control method in reared fish or by otolith daily-increment marking (using chemical imprints or environmental stimuli on the otoliths of experimental fish), but the easiest and most reliable method is the daily-age versus daily-increment control method. When the time that otoliths appear in the ear sacs at late embryonic development to the time of the appearance of the first daily increment is followed continuously, daily-age and daily-increment control can be used to determine whether the otolith increment is a daily increment. It has been suggested that most fish form the first daily increment on the day after hatching (e.g., *Chanos chanos*, *Plecoglossus altivelis*, *Ctenopharyngodon idellus*, *Hypophthalmichthys nobilis*), or when the yolk is close to being absorbed and converted to exogenous nutrients (e.g., *C. harengus*). After the first increment is formed, one increment is formed a day under normal conditions, i.e., the daily increment.

The spacing between the daily increments on fish otoliths varies regularly with the growth and development of fish and environmental conditions. Under natural conditions, the spacing between the first few daily increments usually widens, and then the spacing slightly narrows; after 1 month, the spacing widens as the fish grows and develops and its feeding activity is enhanced. In summer and autumn, when the water temperature is high and the prey is abundant, the distance between the increments widens, and the distance between the increments becomes narrower during the overwintering period; the distance between the rows widens during the growth period of fish, and the distance between the rows becomes narrower during the spawning period of sexual maturity. When fish habitats (e.g., salinity) change, otoliths will leave marker increments that are thicker than the daily rings. *Anguilla rostrata*, *Anguilla japonica*, and

P. altivelis form marker rings on their otoliths when juveniles enter freshwater from estuaries. Marker rings are formed by ecophysiological factors that cause a temporary cessation of daily ring deposition, which usually takes 3–5 days. The shape and size of otoliths vary considerably for different fish species; e.g., the otoliths of the family Staphylinidae are very large, while those of *S. japonicus* and *Trichiurus lepturus* are relatively small.

4.4.4 Methods for Fish Otoliths Daily-Age Determination

(1) Growth Increments for Daily-Age Determination

In 1899, Reibisch discovered that otoliths have an annual ring structure, and in 1971, Pannella was the first to report the existence of a daily growth increment on the otoliths of *Merluccius bilinearis*. After extensive research, it was found that the annual and daily growth increments of otoliths accurately record the life history of fish. At present, otoliths are the most widely used and accurate aging material, and the results obtained by other identification methods are often tested using the results of otolith chronology.

Since the size and shape of otoliths vary from species to species, there are various methods to observe the increments of otoliths. The most common methods are direct observation, fracture cauterization, and embedding and polishing.

The direct observation method is the simplest method and is mostly used for species with small, thin otoliths or for individual juveniles. Since the sedimentation on the otolith surface changes as a fish grows, some annuli cannot be observed from the otolith surface. With cauterization, the opaque areas of an otolith can be turned bright brown, and the transparent areas can be turned dark brown due to protein richness, thus making the annuli easier to observe. In general, the following criteria should be met when using this approach: (1) the fish is older; (2) the otolith is thicker and has sharper edges; (3) the otolith has a blurred opaque area; (4) the otolith has an irregular surface; (5) there are differences in the age readings of different parts of the otolith; and (6) the otolith is incomplete. In addition, otoliths are sometimes abraded after resin embedding to obtain a clearer chronological section. Given the tedious and subjective nature of otolith ring readings, some researchers have investigated the use of computers to automatically count the rings on otolith images. The automatic analysis of annuli is based on the difference in color between the transparent and opaque areas of an otolith, and thus, the transparent or opaque bands are automatically counted. However, there are some difficulties in the process, and they are the following: (1) some false increments may appear in otoliths due to the influence of the fish itself or the environment, which may affect the analysis results; (2) some increments are very close to each other due to the nonlinear regularity of the rings, which is difficult for a computer to recognize; and (3) analyzing the increments requires a high-definition image, so the quality of acquired otolith images is high.

The identification of fish otoliths is a task that requires a high degree of accuracy. The accuracy of age determination using otoliths is highly dependent on fish growth rates, and studies of juvenile *Mallotus villosus* have shown that the faster a fish grows, the more accurate the age determination is. Similarly, age determination using otoliths was found to be more accurate in individuals grown at high prey densities. In addition, the preparation process of the measuring instrument and otolith material has a significant impact on the accuracy of otolith daily-increment identification. For this reason, a number of improvements have been developed for the preparation process of otolith samples.

(2) Otolith Weight for Age Determination

In contrast to individual fish length and other calcified tissue growth patterns, otolith weight in a fish increases throughout its life history. Studies have shown that otolith weight and individual age are closely related, with the average weight of otoliths increasing linearly with age. Therefore, otolith weight is a good indicator of the age of fish, and its application is more promising than other parameters, such as body length and otolith size. A study showed that the age of *C. nasus* was not significantly different from its measured age, so otolith weight can be used as an aid to determine the accuracy of the age of *C. nasus* in the Yangtze River estuary.

Although the determination of otolith weight has the advantages of being a simple operation and having low cost, it also has the following major limitations: (1) Due to long-term overfishing, the age structure of a fish species is relatively homogeneous, and the proportion of younger fish in the catch group is large; thus, there is a lack of information on body length and otolith weight of the older age group, which affects the accuracy of the age determination of the whole stock. The results of studies have shown that otolith weight determination is more effective for younger fish, but the error is larger when applied to older fish. (2) Otolith growth is also influenced by individual growth to some extent. Individuals who grow too fast or too slow in the same population may be overestimated or underestimated when otolith weight is used to estimate their age. (3) Otolith weight can be influenced by the growth environment, which also limits the application of this method. (4) When the number of samples is small, a reliable linear relationship between otolith weight and age cannot be constructed, and therefore, age cannot be identified by this method.

4.5 Lipofuscin and Age Determination of Crustaceans

Age is a necessary component of life history studies of various animals in the ocean, and finned fish deposit specific rings in hard parts (e.g., otoliths, fin rays, vertebrae, scales, etc.), from which their age can be identified by the traditional calcified tissues, but the complex molting process of crustaceans causes the mineralized structures they possess to disappear periodically, making it more difficult to preserve any morphological evidence of age or growth throughout their life spans. This also makes the identification of the age in these animals more problematic.

To address this problem, early scholars mostly used the body length frequency method, which relies on determining the body length modalities of a sample and aliquoting these modalities into grades or supplementary generations. Ideally, the modes can be clearly defined, and the body length frequency distribution is normal, making the analysis relatively simple and straightforward. In practice, however, this is not the case, and the interpretation of body length distribution data is often susceptible to human error. In the absence of other aging methods, modal analysis of body length frequency has been used effectively in many marine vertebrates, but it has been unsuccessful in aging crustaceans, with three main problems: (1) the effects of environmental changes (temperature, salinity) and food quality on growth rates are not taken into account; (2) crustacean growth rates decline with age; and (3) changing growth rates make comparisons between populations unreliable.

The lack of accurate age identification methods makes managing crustaceans in commercial fisheries and conducting resource assessment models more challenging. For a long time, crustacean growth parameters were mostly derived from wild-caught transients, mark-release experiments, and body length frequency methods. In the mid-1980s, some scholars attempted to use lipofuscin in crustacean tissues for age identification with initial success. Since then, the technique has been continuously developed and has now been applied to a variety of crustaceans.

4.5.1 Composition of Lipofuscin and Its Properties

The term "lipofuscin" was introduced by Hueck in 1912 to distinguish this pigment from other age-independent melanins. Lipofuscin is derived from the Greek word "lipo" (oil) and the Latin word "fuscus" (brown), which also implies that early studies provided preliminary insights into the biochemical composition of the substance. However, strictly speaking, the pigment particles associated with aging in cells were first discovered and reported by Hannover in 1842, and these fluorescent age pigments (FAPs), commonly known as lipofuscin, are now considered to be the most obvious age markers in cells.

Compositionally, lipofuscin is an autofluorescent pigment that accumulates in animal tissues as a fluorescent yellow-brown substance. As a byproduct of cellular metabolism, it is composed of lipids (30–70%), proteins (20–50%), carbohydrates (4–7%), and metals (mainly aluminum, calcium, iron, copper, potassium, magnesium, manganese, sodium, and zinc), of which carbon accounts for 57%, hydrogen accounts for 9.3%, nitrogen accounts for 9.1%, and sulfur accounts for 1.6%. This complex chemical composition also gives lipofuscin unique autofluorescence characteristics. The emission spectrum of this compound is broad, and the peak emission wavelength varies with species, muscle sample type, and concentration; however, the maximum excitation and emission wavelengths are usually in the range of 250–450 mm (UV/blue) and sometimes reach 500–640 nm (yellow/orange). However, the degradation of cellular debris results in the production of fluorescent material, which leads to yellow fluorescence when lipofuscin is excited using UV and blue light. This property has also been successfully applied to the discrimination and quantification of lipofuscin particles in many Decapod central nervous system (CNS) tissue sections.

4.5.2 Age Determination Using Lipofuscin

In studying the biology of crustacean fisheries, accurate aging techniques can be of great value on several levels. Although lipofuscin was first described as a xanthophyll in nerves in 1842 and was shown to have age-discriminatory qualities, it was not until 1886 that researchers discovered a link between it and age. Until the early 1980s, information about the compound was largely available only in health and pharmaceutical literature. However, since then, lipofuscin has attracted the interest of ecologists because of its use in aging invertebrates.

Because lipofuscin concentrations increase with biological age, it has also been used as a biomarker of age in many invertebrate aging studies when no relationship between morphological characteristics and age can be established, and using lipofuscin concentrations has repeatedly shown advantages over using body length or body weight aging techniques. For example, in *Cherax quadricarinatus*, the mean age prediction error for the 180–780-day age group was 16.65% for the lipofuscin aging method, while the error of the carapace length frequency method reached 32.45%. For *Homarus gammarus* aged 4.3–8.4 days, there was no significant relationship between carapace length and age in small individual marker samples, whereas relatively reliable ages were obtained for these lobsters using the calibrated lipofuscin aging method.

(1) Principal Characteristics of Lipofuscin

Lipofuscin has three main characteristics: (1) it is contained in intracellular lysosomes; (2) it emits yellow autofluorescence when excited by UV or blue light; and (3) it accumulates with age in postmitotic tissues. These features and their unique biochemical and morphological characteristics have led researchers to consider lipofuscin as an "age marker" (Zhu and Song 2016).

Lipofuscin accumulates with age in a wide range of organisms, including mammals, fish, insects, crustaceans, bivalves, and nematodes, and is widely present in older individuals. Usually, the suitability of a tissue for lipofuscin aging is largely determined by its cellular dynamics and the activity of extracellular solute remnant vesicles. Some neural and glial parts of the brain can maintain and accumulate lipofuscin with age throughout life, which also provides an ideal target for lipofuscin aging.

As a pigment that accumulates in postmitotic tissues such as the brain and optic stalk, lipofuscin accumulates with age, a process that can be revealed by the deutocerebral optic lobe and the protocerebral medulla terminalis. Although the amount of lipofuscin in the optic medulla terminalis of the otic stalk is lower than that in the optic lobe, the optic medulla terminalis is easy to sample and does not kill the sample, making the application of this method more promising.

Lipofuscin content can be quantified by observing tissue sections and analyzing these sections by fluorometric spectroscopy, and the results are usually expressed as a percentage of area or volume. Ettershank (1984) first determined the age of *Euphausia superba* using lipofuscin by solvent extraction and fluorometric spectroscopy. In 1990, Sheehy used brain tissue sections to determine lipofuscin content directly by histological methods (Sheehy 1990) and claimed that the method was applicable to the age determination of many crustaceans. Studies in *C. quadricarinatus* showed a higher correlation between age and lipofuscin content than between lipofuscin and body length (Sheehy 1990); however, the study was based on organisms cultured in the laboratory, and changes in the field environment may have affected the results. There are interspecific differences in the extent of lipofuscin accumulation with age, even depending on the environment to which the same species is exposed. The process is closer to a reflection of the physiological age of an organism and the growth from individual metabolism than to a process simply identified with time. Therefore, the lipofuscin content measured in crustacean brains is only an approximate of their physiological age, rather than their actual age. In addition, calibration of the organism and the environment corresponding to the known biological age is required.

(2) Quantification of Lipofuscin

Although there are multiple methods to quantify the content of crustacean lipofuscin, in general, they can be divided into two main categories: the first method is the histomorphometric method, which involves sectioning and staining nerve tissue, followed by a combination of fluorescence or confocal microscopy and digital analysis, using fluorescence (confocal) microscopy to count and quantify the pigments containing particles, and then, a morphometric analysis of lipofuscin particles in tissue sections using digital techniques is conducted; the second method is the solvent extraction method, in which the pigments are extracted using organic solvents, and then, the fluorescence content is determined by spectrophotometric analysis.

The histomorphometric method usually involves first cutting postmitotic tissue containing age pigments from animal tissue and then observing sections of that tissue using confocal fluorescence microscopy. When the observed tissue is exposed to the optimal excitation wavelength, the lipofuscin particles become autofluorescent. The fluorescent particle images are then digitized to determine the average volume or area occupied by the lipofuscin relative to the entire area of the tissue sample. The larger the average volume or area occupied by the lipofuscin, the higher its age pigment content is, and thus, the greater the physiological age of the organism is. This technique has been used successively to determine the age or population age structure of a wide range of crustaceans. Generally, the histomorphometric method is more laborious and requires substantial sample processing time. Although the method is accurate in estimating age and usually predicts the age of organisms to within a few months, the time-consuming nature of the method and the small sample size used also reduce its statistical effectiveness. The slicing process makes it difficult to fix the structure of tiny tissue blocks in paraffin, and this is a key issue that needs to be addressed to be able to produce quantifiable results from this method. In addition, given the difficulty of obtaining organisms of known ages under field conditions, calibration of the relationship between age and lipofuscin content was not better resolved in most cases, although the mark-release test may partially solve the problem.

Early solvent extraction methods included ethylene-based biochemical procedures, in which age pigments were extracted from tissues using Folch's method. Later, the method was adapted by some scholars. After lipofuscin extraction, the fluorescence intensity is measured and quantified by spectrophotometry. However, there are still methodological problems with this process, with the main problems being that the fluorescence content obtained from tissue extracts is dependent on the means of handling and preserving the samples and that other age-related autofluorescence products contained in the extract can interfere with the quantification results. Additionally, determining what fluorescent material is extracted in relation to age involves calibration and verification of the method. The best current approach is to follow the accumulation of fluorescent products over time by means of laboratory transient experiments. The results of the study also showed that lipofuscin was easily extracted using a chloroform-methanol mixture and that there was a correlation between the concentration of fluorescence extracted and the age of the organism. Since that study, several researchers have used this method to successfully estimate the age of other invertebrates.

(3) Calibration of Lipofuscin

A central tenet of accurate age identification is the need to calibrate the method using organisms of known age so that the rate of oxidation product accumulation in these organisms can be tracked. There are three common methods used for lipofuscin age identification calibration: (1) transient rearing of individuals of known age in the field with temperature conditions similar to those at the field site of interest, (2) analyzing lipofuscin concentrations measured in situ in field individuals using frequency histograms of their generations, and (3) labeling juveniles of known ages and recapturing them after a number of hours. All three of these methods have advantages and disadvantages.

New methods for crustacean age determination have emerged over the past few decades. Since 1886, lipofuscin has been extensively studied in vertebrates and invertebrates given its role in animal aging, but the method has not been more widely used in fisheries because of uncertainties in the quantification and calibration of lipofuscin.

4.6 Cephalopod Beaks and Its Age Determination

The study of age and growth is an important part of research on the biology and ecology of cephalopod fisheries. In recent years, the cephalopod beaks have received increasing attention from many scholars because they are individually larger than otoliths and have the advantages of being easy to extract and easy to grind and are therefore considered to be one of the most reliable hard tissues used to identify the age of cephalopods.

4.6.1 Structure of the Beak

Beaks are the main tearing and feeding organ of cephalopods and are mainly used to tear food. Beaks are composed of two parts, the upper beak and the lower beak, with a mosaic pattern of the lower beak covering the upper beak, in contrast to the mosaic pattern of the beak. The upper and lower beaks of cephalopods are similar in structure and consist of five main parts: rostrum, wing, lateral wall, hood, and spine (Fig. 4.7). The main length parameters are hood length (HL), which is the length from the tip of the rostrum to the end of the posterior margin of the hood; crest length (CL), which is the length from the tip of the rostrum to the end of the

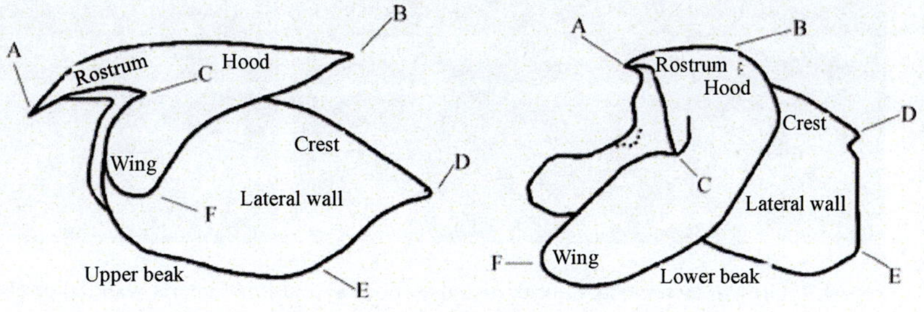

Fig. 4.7 Schematic diagram of the morphological parameters of a beak (Liu et al. 2017). A–B Hood length; A–C rostral length; A–D crest length; A–E lateral wall length; and C–F wing length

posterior margin of the crest; rostrum length (RL), which is the length from the tip of the rostrum to the end of the beak angle; lateral wall length (LWL), which is the length from the tip of the rostrum to the end of the beak angle; and wing length (WL), which is the length from the angle of the beak to the end of the anterior edge of the wing (Fig. 4.7).

Similar to otoliths, inner shells, eye crystals, and other hard tissues, cephalopod beaks have obvious growth pattern structures. The growth patterns on the surface of the cephalic, crest, lateral wall, and wing parts of the cephalic beaks are obvious and visible to the naked eye in the form of fluctuating strips, while the growth patterns on the rostrum need to be cut and ground to be visible. Therefore, methods to observe growth patterns of the beak are divided into surface and internal growth pattern observation methods: (1) the surface growth pattern direct observation method involves cutting the beak longitudinally along the posterior edge of the hood toward the tip of the rostrum, and then, the growth pattern on the inner surface of the lateral wall with clear lines is observed directly under the dissecting microscope (Fig. 4.8a); (2) the internal growth pattern observation method involves cutting the beak longitudinally along the posterior edge of the hood toward the tip of the rostrum, and then, it is encrusted, ground, polished, and observed under the microscope (Fig. 4.8b).

4.6.2 Growth Increment in Beak

Clarke first reported the structure of growth lines in cephalopod beaks in 1962, followed by a special study of the growth increments in the beaks of *Moroteuthis ingens* in 1965 and similar structures in cephalopod beaks by Nixon in 1973 and Smale et al. in 1993. In 1998, Raya and Hernández-González first hypothesized that the regular growth deposited inside the beaks' rostrum of *Octopus vulgaris* might be related to their age, and it was not until 2001 that Hernández-López experimentally confirmed the daily periodicity of the growth pattern of *Octopus vulgaris* beaks, which conformed to a "one-day cycle."

Based on the observation of the growth increment inside the upper or lower rostrum of *Octopus maya*, a significant linear relationship ($P < 0.0001$) was found between the actual day age and the number of beak growth increments, with slopes and correlation coefficients R of 0.9967 and 0.9945, respectively, with both close to 1. This result further confirms that the beak growth increments of *O. maya* have a significant daily periodicity. By observing the growth increments of five ommastrephids, including *Illex argentinus*, *Todarodes pacificus*, *D. gigas*, *Ommastrephes bartramii*, and *Sthenoteuthis oualaniensis*, hatched in captivity, it was found that the cuticular frontal deposition of all five cephalopods conformed to a one-day cycle,

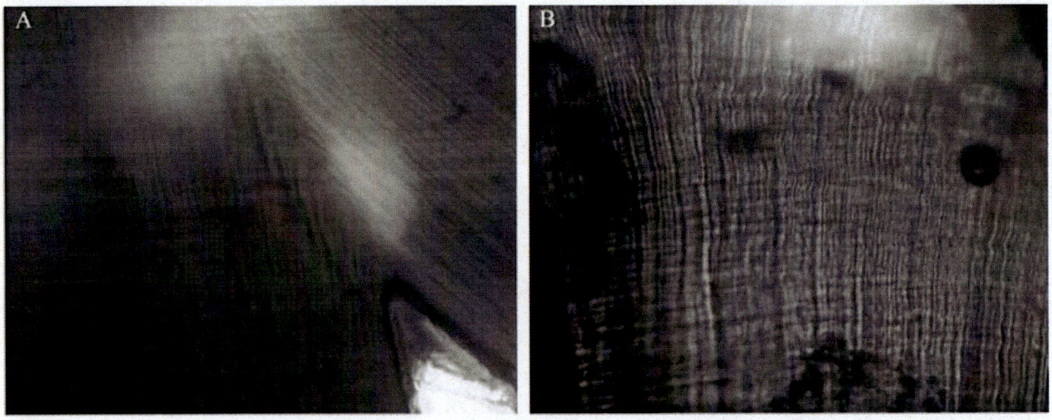

Fig. 4.8 Growth increments in beak rostral section (**a**) and inner lateral wall (**b**) of *Dosidicus gigas* (Hu 2016)

except for *S. oualaniensis*. Although the daily periodicity of the growth pattern of the beak of most cephalopods has not been confirmed, there seems to be a consensus among cephalopod scholars that the growth pattern of the beaks conforms to a "one-day" growth pattern.

However, cephalopod beak growth increment observations are also subject to various sources of error; for example, the age estimated from the rostral growth increment of beaks is often younger than the actual age because the rostral tips of field-collected cephalopod beak frontal samples are often corroded or damaged by feeding. In addition, the timing of beak formation varies among cephalopods; for example, in the analysis of the cuticular fronts of captive cephalopods (*I. argentinus*, *T. pacificus*, *D. gigas*, *O. bartramii*, and *S. oualaniensis*), it was found that the first growth pattern of the cuticular fronts began to form on the first day after hatching in *D. gigas* and *O. bartramii*, while the first growth pattern of the cuticular fronts began to form on the second day after spawning in *I. argentinus* and *T. pacificus*.

4.6.3 Preparation of Beaks and Daily-Age Determination

4.6.3.1 Extraction of Beaks

Buccal masses are removed from fresh cephalopod samples, placed in a glass container, labeled, and left at room temperature for 24 h to allow for muscle decay, and then, the beaks can be extracted with forceps. Frozen samples can be extracted directly from the upper and lower beaks with forceps after they have thawed.

4.6.3.2 Preservation of Beaks

After the beaks are extracted, they are placed in water to clear off the surface mucus, soaked in a pepsin solution for 2 days to remove any residual organic matter on their surfaces, and finally placed in a 70% alcohol solution to prevent dehydration.

4.6.3.3 Preparation of Beaks

1) Preparation of Rostrum Sections of Beaks

The preserved upper beak is removed and cut in half longitudinally along the top of the rostrum to the posterior edge of the hood using a small handheld cutter with a 0.3 mm blade (Fig. 4.9),

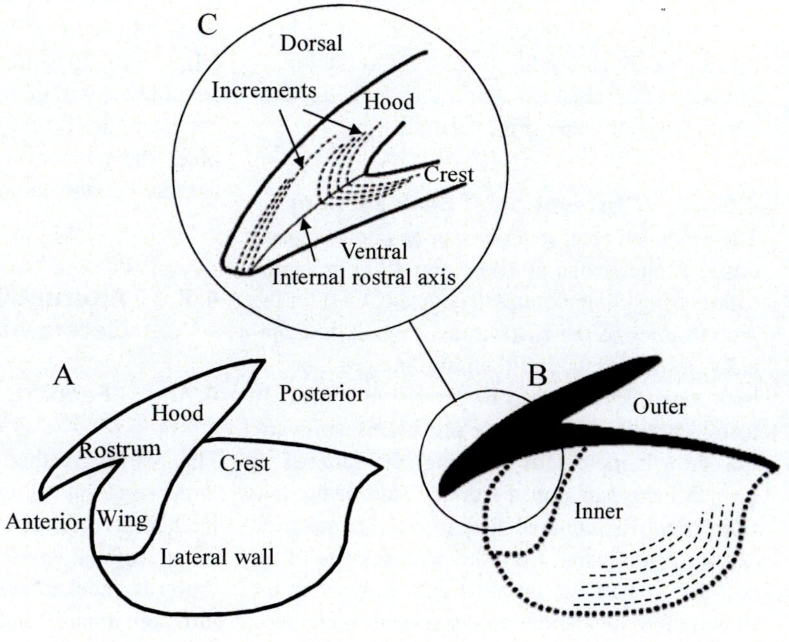

Fig. 4.9 Schematic diagram of the upper beak of squid (Liu et al. 2017). (**a**) Composition of each part of upper beak; (**b**) half of the beak cropped rostrally along the crest and posterior margin of the hood; black portion indicates the cross section; (**c**) enlarged cross section of rostrum, showing regular growth increments

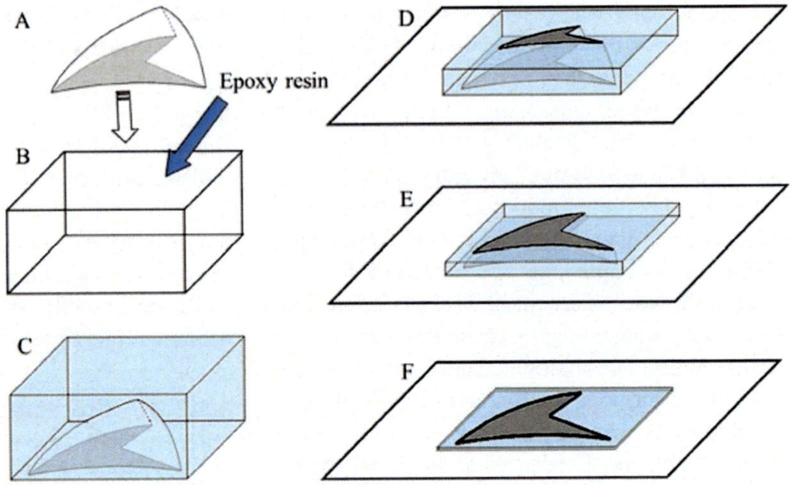

Fig. 4.10 Flowchart of the production of rostrum sections (Liu et al. 2017). (**a**) Cut rostrum; (**b**) plastic abrasive; (**c**) resin-embedded rostrum; (**d–f**) cut resin blocks glued to slides and ground to the central surface

and the rostrum sagittal section (RSS) of one half of the beak is cut with scissors (Fig. 4.10a); the cut side of the rostrum section is laid flat face down on a hardened resin block that is then cut into 2–3 mm slices and glued to the slides with strong, hot adhesive (Fig. 4.10d); the slices are ground to the central surface with 240, 600, 1200, and 2000 grit waterproof sandpaper, and finally with 0.05 μm alumina agent to polish the grinding surface (Fig 4.10e, f).

2) Preparation of Lateral Wall Sections of Beaks

The preserved upper beaks are, cut in half longitudinally along the rostrum tip to the posterior edge of the crest using a small handheld cutter with a 0.3 mm blade (Fig. 4.9).

4.6.3.4 Observation of Beak Sections

The produced beak sections can be photographed under a microscope at 100× and 400× magnification using charge-coupled device (CCD) for the growth lines of the rostrum as a whole and apically, respectively, and then, the images taken at both magnifications can be processed separately using Photoshop 7.0 image processing software. For growth increment counting, the number of growth increments on the dorsal side of the rostrum (hood) is counted first for images taken at 100× magnification from the posterior end of the rostrum toward the anterior end, always in the direction perpendicular to the growth increment, up to the dorsal edge of the rostrum, and then, this process can be continued for images taken at 400× magnification, with the number of growth increments in the marginal beak being derived from the width of the adjacent growth increment (Fig. 4.11).

In addition, the experience of those counting the growth increments is critical to the accuracy of the day counts, and the reproducibility of the count results can be used as a test of accuracy. Therefore, to improve the accuracy of counts, the results of trained counters can be compared, and the results are credible if there are no significant differences. In general, the critical criterion for credible growth increment counts in age identification studies is that the counts are repeated three times independently and the difference between the three counts is no more than 10%.

4.7 Alternative Methods for Age Determination of Fish

4.7.1 Feeding Method

The rearing method is the most primitive and direct method to check the age of fish; this method involves fish of known ages being reared in an artificial environment, and the growth condition is checked periodically to study the structure of annuli and duration of annual ring

Fig. 4.11 Schematic diagram of the growth increment of within rostrum of squid (Hu 2016)

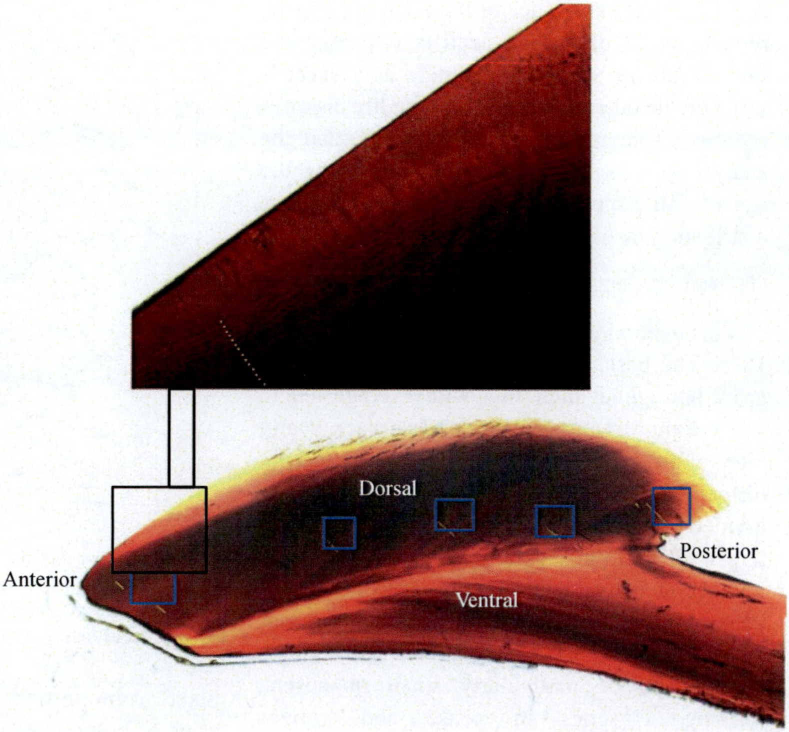

formation and to further explore the reasons for annual ring formation annuli and the influence of environmental factors on the growth of fish. However, some scholars have pointed out that the accuracy of age determination by the rearing method is not high because the growth rate of fish in artificial rearing environments is generally slow. At the same time, the breeding environment cannot sufficiently mimic the real natural environment, the formation of the annual ring is greatly influenced by the environment, and an artificial environment is likely to lead to the production of artificial ring patterns, thus causing bias in the calibration process. However, since otolith daily rings are less affected by the environment, some studies have shown that the daily rings obtained by artificial rearing provide similar results obtained under natural conditions; therefore, the rearing method is more valuable for daily ring identification.

4.7.2 Mark-Release Method

The mark-release method is one of the most effective methods of age determination. It determines the accuracy of age determination based on bone structure and involves marking and releasing individuals of known age, then analyzing the bone structure of recaptured individuals for their age structures, and comparing the results with the true age. However, this method is not suitable for long-lived fish because the recapture rate of marked fish decreases with time. In addition, this method is suitable for species that can be reared, as a rearing process is usually required prior to release.

4.7.3 Length Frequency Method

The use of life history is the most accurate method to determine the age of fish, but it also has some drawbacks: on the one hand, the life history determination requires some skills, which mainly depend on the experience of the scientist, and the probability of different scientists obtaining the same result for the same material rarely exceeds 90%; on the other hand, determining life histories is time-consuming and labor-intensive, and the analysis cost is high. Therefore, estimating the age of fish populations based on the frequency distribution of fish lengths is also of interest.

(1) Natural Length Distribution Curve Method

This method was first proposed by Peterson in 1895. The basic principle is that individual fish grow throughout their life, with average length and weight varying by one level at each yearly interval. In the same population, there are usually different age groups so that all individuals can be divided into several length groups. After the length data of the samples are measured, the number of fish in each length group is plotted; there will like be some successive length groups that are particularly numerous, while some length groups will be particularly small or absent, forming a series of peaks and troughs (Fig. 4.12). Each peak represents an age group, and the length group of each peak represents the body length range of that age group. Under conditions where gear is not selective, the peaks in the length group distribution curve are generally sequentially lower, as in the case of eastern Icelandic cod (Fig. 4.13).

However, there are some limitations to this approach. First, each gear is somewhat selective in its catch, and it is difficult to include all age groups in a catch at the same time. Second, the seasonal periods of fish in each fishery are not mixed in proportion to their natural numbers in terms of body length or age. As older fish enter their senescence periods, growth slows or even stops, and there is inevitably an overlap in length distribution; thus, it is not easy to determine the age of older fish from a length distribution curve. Finally, during the growth and development of fish, whether prey is abundant or not and whether the water temperature is suitable or not directly affect the size of fish. Therefore, it is recommended that the natural length distribution curve method be used in conjunction with other identification methods.

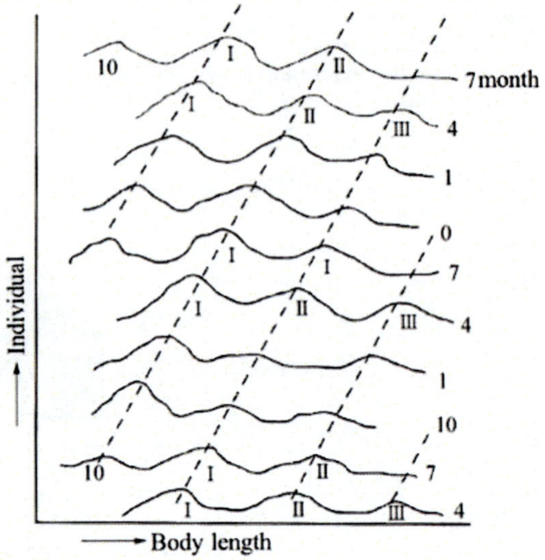

Fig. 4.12 Distribution of continuous specimen age groups (Chen 2014; Chen and Liu 2017). 0 for current age fish, I for 1-year-old fish, II for 2-year-old fish, and III for 3-year-old fish

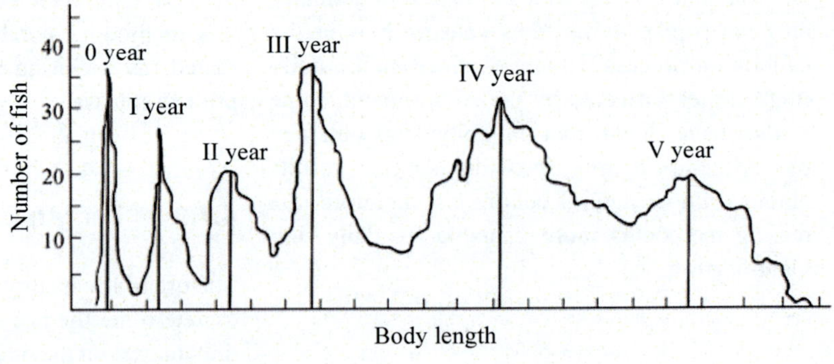

Fig. 4.13 Length distribution curves for cod in eastern Iceland (Chen 2014; Chen and Liu 2017)

Table 4.1 Dominant age groups of mackerel off Yantai (Chen 2014; Chen and Liu 2017)

Year	1953	1954	1955	1956	1957	1958
Dominant age group	IV	V	VI	VI–VII	VIII	X

(2) Graham M. Dominant Body Length Method

By analyzing the composition of catches every year and using the growth of the dominant length group, we can determine whether the circulus on scales (or on bones) grows once a year; for example, if a certain fish is in a particularly large body length group with scales of 20 cm in the previous year and the number of circulus on the scales of a 30 cm body length fish this year is one more than last year, then the circulus on the scales should be true annulus. Therefore, the annual growth cycles of fish form repeated circulus on the skeleton, scales, and other morphology. Based on the appearance of this circulus, the age of a fish can be determined. A prerequisite for the use of this method is the presence of a dominant age group (dominant body length group) in a fish population; i.e., the number of fluctuations in each generation of this fish resource is not too disparate. According to the type of spawning groups classified by T.H. Monastery, i.e., species belonging to the second type, complementary groups often dominate over the remaining group. For example, the mackerel off Yantai, China, is of this type (Table 4.1).

In another study, based on information on the composition of Irish cod otolith rosettes, the rosette group that dominated for several consecutive years (Fig. 4.14) was used to determine age (black bars are the dominant age group).

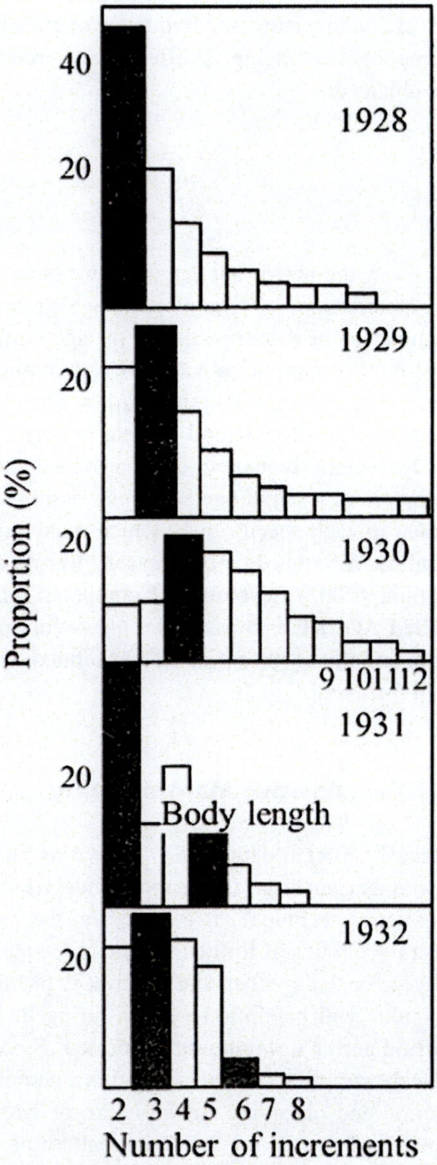

Fig. 4.14 Age determination of cod based on body length distribution (Chen 2014; Chen and Liu 2017)

(3) Age-Body Length Conversion Table Method

The age-body length conversion table method is a widely used method of age determination. As early as 1934, Fridrikson proposed the use of body length frequency data to obtain age frequencies. After decades of development and calibration, it has become a relatively complete identification method. The main implementation method is as follows: first, secondary sampling from a large fish sample is conducted using otoliths or other relatively reliable methods to identify fish ages, the age and body length data of each fish are recorded, and they are grouped together. With length groups as rows and age groups as columns, the number of fish in each group can be listed as follows:

	A1	A2
I1	6	2
I2	3	3
I3	1	4

After conversion, the frequency of each length component belonging to different age groups can be obtained:

	A1	A2
I1	0.75	0.25
I2	0.50	0.50
I3	0.20	0.80

From the above age-length conversion table, the percentages of fish of each age group in a sample can be determined, and the age frequency distribution can be obtained by summing the percentages of each length group by age group.

However, this method of determining age by body length frequency only provides the age structure of a sample and does not assign an age value to each specific fish, which makes further analysis difficult. For this reason, Isermann and Knight (2005) developed a computer program called AGEKEY to assign an age value to each fish, and this method was further refined by Ogle and Isermann (2017).

4.7.4 Relative Margin Dating

Relative margin dating, also known as marginal increment analysis, is the most widely used age correction technique. It is based on the premise that if a circulus is formed within a time period of 1 year or 1 day, then the outermost part of the circulus will continue to grow during that time period until a new circulus is formed. By continuously sampling fish species within a certain time period and observing the structure of circuli on their hard tissues, the formation pattern of circuli can be determined, and the technique of age identification by circuli can be verified. However, this technique is difficult to perform, mainly because the edges of the material to be observed are usually very thin, and these types of observations are disturbed by refracted light. In addition, some studies have found that marginal incremental analysis is more reliable for juvenile and fast-growing fish, but when applied to older fish, it produces large errors. Additionally, environmental factors such as temperature have a strong influence on marginal increment analysis, so it is believed that marginal increment analysis cannot be used for fish grown under severe temperature changes. In addition, the error of otolith marginal increment analysis also originates from three main sources: (1) an insufficient sample size, (2) a long data collection period, and (3) a long duration of fish birth.

A relative margin is calculated by taking a certain number of specimens from a catch, monthly during an annual period, and observing the changes in the growth of the rings on the scales at the edge of the scales to determine the cycle and time of scale formation to determine the age of the fish, and there are two methods to measure the growth of scale edges.

The first method is to calculate the ratio of scale edge growth to scale length.

$$K = \frac{R - r_n}{R}$$

where r_n is the length of each circulus, R is the scale length, and K is the relative marginal growth value. The disadvantage of this equation is that the denominator value R becomes larger with increasing age so that the ratio also decreases in higher age groups.

The second calculation method is based on the change in ratio K of the magnitude of scale edge growth $(R - r_n)$ to the distance $r_n - r_{n-1}$ between the last two circuli of that scale as an indicator for determining the cycle and time of annulus formation. It is calculated by the following equation:

$$K = \frac{R - r_n}{r_n - r_{n-1}}$$

where K is the edge growth value, R is the distance from the center to the edge, R-r_n is the distance from the edge to the penultimate first round, and $r_n - r_{n-1}$ is the distance between the penultimate first and second rounds. The wider the edge, the larger the K value is; conversely, the smaller the K value is. At the beginning of the

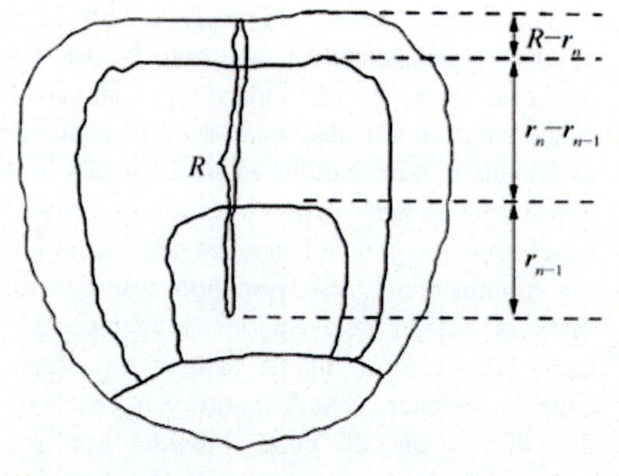

Fig. 4.15 Schematic diagram for measuring the magnitude of scale edge growth (Chen 2014; Chen and Liu 2017)

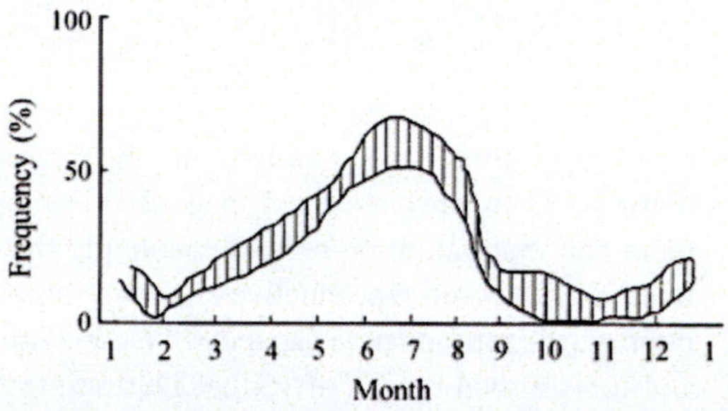

Fig. 4.16 Frequency map of the occurrence of annulus on the scales of the East China Sea white gourami (Chen 2014; Chen and Liu 2017)

formation of a new circulus, the K value is extremely small, close to 0. When the K value gradually increases and the edge amplitude approaches the width between two circuli, a new circulus is about to appear (Fig. 4.15).

An analysis was carried out on the Chinese East China Sea white gourami, whose first annuli form exactly at the edge of the scales, and the frequency curve is shown in Fig. 4.16. In the lower part of the curve, the percentage of the first annuli formed at the edge of the scales is indicated. In the upper part of the curve, the percentage of rings not formed is indicated, and the middle line indicates the percentage of the likely annual ring composition. As seen in Fig. 4.16, the period during which the annuli are formed is from June to August, indicating that there is only one peak in the year, and therefore, only one annual ring can be formed; thus, the first annual ring can be identified.

4.7.5 Isotope Calibration Method

Kastelle et al. (2000) first proposed in 1982 that the ratio of the isotope $^{210}Pb/^{226}Ra$ in fish otoliths could be used to verify an age determination based on otoliths. The principle of this method is that the isotope decays into other elements after a certain period of time after entering an otolith, and the half-life of each element is known; thus, the ratio of these two elements can be used to determine the amount of time the fish has lived. Campana et al. (1990) proposed the following three assumptions for the $^{210}Pb/^{226}Ra$ isotope calibration method: (1) the otolith is a closed environment and is not affected by external radioisotopes during the decay process; (2) the initial $^{210}Pb/^{226}Ra$ ratio should be below 1, close to 0; and (3) the radioactivity of the otolith absorbed by a fish remains constant throughout its life history. Of these, the first assumption is the most difficult to guarantee.

Age determination techniques for fish have been developed over hundreds of years and have formed a relatively strong identification system. However, various age identification techniques have shortcomings. The main limitations are as follows: (1) the accuracy of some determination methods is not high, and accuracy is influenced by the subjective factors of the scientist implementing the methods; this is especially important when age is identified by hard tissue rings, so the experience and skill levels of the scientist must be high. (2) Most identification methods are tedious and time-consuming. Since age identification work is less automated, most work is conducted manually, so it often takes a long time. Although computer automation technology is gradually being applied to age identification, there are still many shortcomings, and it will take time for it to become a mainstream identification technology. (3) Fish age identification is not standardized. Different scientists often use different fish tissues to age a species, and even if the same tissue structure is used, the processing techniques and observation methods vary greatly.

4.8 Fish Growth and Measurement Methods

4.8.1 Patterns of Fish Growth and Their General Characteristics

(1) Division of Fish Growth Stages

The growth stages of fish can be divided into three periods: in the first stage, fish have not reached gonadal maturity, growth fluctuations are very severe, and sufficient prey, rapid growth, and starvation are the main factors affecting growth; in the second stage, fish are in the period of sexual maturity, all the body stores are mostly transformed into reproductive products, spawning or sperm production activities are carried out every year in the breeding season, growth is stable from year to year, and the growth rate does not change. The third stage is senescence, when metabolism decreases, growth is slow, and body length and weight increase very slowly with increasing age until death.

(2) rowth Regulation of Immature Fish

The growth characteristics of fish vary by developmental stage. The most rapid growth of fish length usually occurs after the end of the juvenile stage and before gonads are fully mature, i.e., during the juvenile period. This is the time when foraged prey, most of whose nutrients can be converted into substances in the body, is growing uninterruptedly and when the fish are not accumulating and storing more substances, especially fat. For example, juvenile and young freshwater salmon and trout are in a vigorous feeding phase during all three seasons of the year, slowing feeding only in winter. After entering spring, water temperatures warm slightly, and foraging is again intense to maintain body length and weight growing rapidly. After years of migrating and foraging, gonads mature; once the gonads are mature, body length growth slows.

The rapid increase in body length occurs before sexual maturity to protect fish from predators and reduce mortality. Fish growth characteristics are also related to reproduction, as gonads mature rapidly when fish reach a certain size and length. In prey-rich waters, growth is very rapid, and gonads can reach maturity in just a few months or a year; in prey-poor waters, sexual maturity will be delayed for a longer period of time. For example, in the case of the offshore Chinese croaker, sexual maturity is reached in the third year of good growth, but some individuals delay sexual maturity for 5–6 years. Therefore, the sexual maturity of fish is closely related to body length growth but not directly related to age.

(3) Regulation of Growth During Sexual Maturity

During gonadal maturation in fish, most nutrients from prey are used in gonad development and maturation process to enable the metabolic activities of reproductive products and to improve the reproductive capacity of a population and the survival rate of offspring. During sexual maturity, the level of fecundity is closely related to the length and weight of fish. Usually, individuals with long body lengths and high body weights generally have higher egg carrying capacities and greater ability to reproduce offspring; conversely, fish with short body lengths and low body weights have lower egg carrying capacities and less ability to reproduce the next generation. Most fish have irreversible growth characteristics. An increase in length and weight is also related to seasonal rhythms. Most fish that inhabit temperate waters are characterized by overwintering habits, with the greatest growth occurring in the fall and the least growth occurring during the spawning period, regardless of the size of the individual or the density of the school. In comparison to those in temperate zones, fish in tropical zones show less significant variation in growth during the different seasons of the year.

(4) Growth Regulation During Aging

Senescence in fish is the slowing or stagnation of normal metabolic processes in fish, during which prey is used only to sustain life activities, and body length and weight increase more slowly. The fecundity of fish during senescence decreases, especially when gonads are underdeveloped, are atrophied, or have low fertilization rates.

The fish that enter senescence vary by species, and the same fish inhabiting different waters also vary. Individual fish of the same species also have different senescence periods. Fish that sexually mature earlier usually have a shorter life span, while those that sexually mature later may have an increased number of reproductive spawning and a slower aging period; e.g., a fast-growing fish that matures sexually at 3+ years of age has its last reproductive spawning activity at 7 years of age, while individuals who mature sexually at 5+ years of age still participate in reproductive activities at 12 years of age.

4.8.2 Main Factors Affecting Fish Growth

Fish growth is influenced by various factors, which can generally be divided into internal and external factors. Internal factors refer mainly to physiological and genetic aspects, and these are the main elements of biological research. For example, in fish, generally, female individuals are larger than males, and males generally mature before females and grow earlier and at slower rates. Exogenous factors refer to the effect of external environmental factors on growth.

The external environment in which fish live can be subdivided into biological and nonbiological aspects. Biological aspects of the external environment are mainly reflected in the predator-prey relationship, i.e., predation by aggressive animals. Fish grow faster before sexual maturity to defend themselves from predation by aggressive animals. Prey organisms are the source of energy for fish growth, the amount of which directly affects growth and development. Under optimum water temperature conditions, adequate prey supply is a key factor in promoting fish growth, i.e., the quality and quantity of bait. If prey organisms are scarce and low quality, then the growth and gonadal development of fish are seriously affected. In natural marine areas, the abundance of prey varies due to seasons and regions, and in farming ponds, prey feeding is very important and is the most critical factor in the success or failure of fish farming.

Abiotic aspects of the external environment include temperature, salinity, dissolved oxygen, light, and other factors. Fish have certain requirements for temperature, regardless of the stage of life, and temperature will directly or indirectly affect the growth of fish. In general, within the appropriate range of water temperatures, the higher the temperature is, the faster the growth and metabolism of fish. Each kind of fish has its most suitable range of water temperature. Under these temperature conditions, fish metabolism is the most active and most vigorous with the strongest physiological response capacity, and the fish body rapidly grows and develops. If the water temperature is too high or too low, then it can affect the development of gonads and the survival rate of eggs or sperm or even cause death. For example, fertilized salmon eggs hatch at water temperatures of 12 °C, and parents can tolerate water temperature changes of 20 °C.

There are seasonal variations in the growth rates of fish, and the growth status of fish varies from one latitude to another. The growth rates of fish born in different generations significantly differ, so there are abundant and nonabundant years.

4.8.3 Methods for Measuring Fish Growth

4.8.3.1 Direct Measurement Method

Based on the life history and growth information obtained for each sample of each catch, the average lengths of each age group can be grouped together to calculate directly observed fish growth values and thus the actual length of growth per year. Provided that growth rates at age do not differ significantly between generations and that individual age groups consist of random samples, the average lengths at these ages can be used to directly estimate the year-to-year growth rate of fish.

Studies have shown that fish increase rapidly in absolute values of length, and these increases are positively correlated with increases in absolute values of body weight during the first stages of life. Subsequently, increases in length and weight slow with age. Some species can spend 3–5 years in the initial stages, some 8–9 years, or even a little longer. This means that in relation to fish longevity, the initial growth phase is completed within 1–2 years for short-lived fish, and for long-lived fish, the initial growth phase can be extended by a number of years as appropriate, with most economic fish completing their length growth or weight gain within their first 2–3 years.

The best time to obtain samples for direct measurements is during the breeding season, winter or the season when new annuli are formed. The advantage of direct dating is that this approach is the closest to the actual conditions and reflects the true nature of the subject. The disadvantage is that the data obtained at one time cannot contain all the age specimens needed. Specimens obtained from different fisheries may have different growth rates and do not reflect the growth of the same generation of fish well, and the data obtained reflect only the growth of different generations of fish at the same rate.

4.8.3.2 Back-Calculation of Length

The principle of performing back-calculations of fish length is that the growth of fish is very closely related to the type, quantity, and size of its prey, the water temperature, and the quality and density of its habitat. Fish grow rapidly in waters where prey is extremely abundant and habitat conditions are suitable.

Walter (1901) studied *C. carpio* growth and found that the circulus of their scales was proportional to the length of the fish (Fig. 4.17). In the same year, the Norwegian scholars Lien (Lea) and Dahl (Dahl) also obtained this result and concluded that the growth of fish scales increases with age and that the length of scales is proportional to the length of the fish. Their equation is the following:

$$L_t = \frac{r_t}{R} \times L_0$$

where L_t represents the length of the fish in previous years, L_0 represents the measured length at the time of capture, r_t represents the length of the scales in the year corresponding to L_t, and R represents the length of the scales at the time of capture.

Since those studies, many scholars have developed a series of equations after continuous research. At the same time, it was found that there was a certain error between the above equations and the actual measurement when

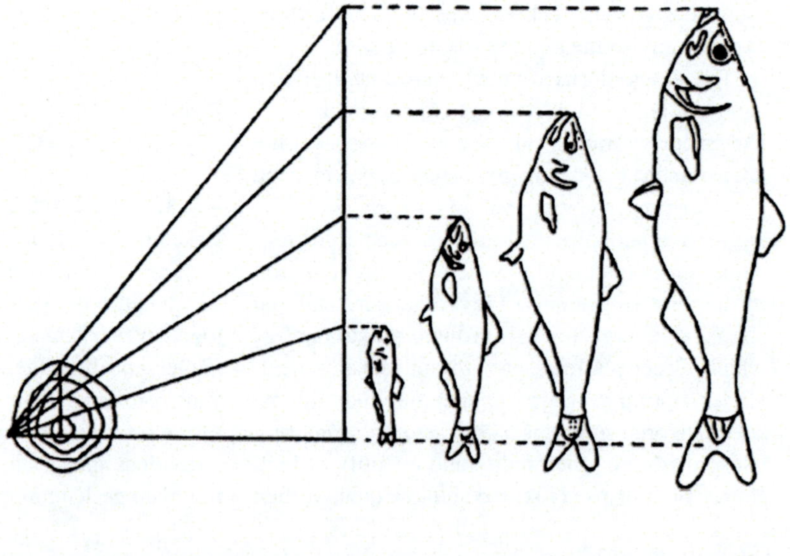

Fig. 4.17 Interrelationship between fish growth and scale growth (Chen 2014; Chen and Liu 2017)

projecting fish growth; i.e., the length of fish in the projection was smaller than the value directly measured. This error is particularly pronounced in older fish, a phenomenon known as the Rosa Lee phenomenon. The shortcoming of this method is that it does not take into account the growth characteristics of fish length and scale length because the scales appear after a fish have grown to a certain length, not just after hatching. For this reason, in 1920, Rosa Lee revised the equation to the following:

$$L_t = \frac{r_t}{R} \times (L - a) + a$$

where a represents the length of the fish when scales begin to appear.

The above equation indicates that scale growth is linearly related to the growth of a fish as in the following equation:

$$L = a + bR$$

where a and b are constants, a is biologically significant and the body length at the time of scale appearance, and b corresponds to the body length per unit of scale.

Further research later concluded that some fish scales do not grow in a linear relationship with body length but conform to other growth patterns, such as power exponential, parabolic, and hyperbolic.

4.8.3.3 Calculation of Fish Growth Rate Types and Growth Indices

Growth rate calculations for fish typically include the following:

1) Absolute growth rate (or rate of weight gain) in a given year: L_2 -L_1 or W_2 -W_1.
2) Relative growth rate: (L2-L1)/L1 or (W2-W1)/W1, usually calculated as a percentage.
3) Instantaneous growth rate: lnL2-lnL1 or lnW2-lnW1, representing various forms of typical population growth curves.
4) Relative growth rate: expressed in terms of logarithm of growth

$$C_e = \frac{lgL_2 - lgL_1}{0.4343(t_2 - t_1)}$$

where 0.4343 is the coefficient of conversion of the natural logarithm to the general logarithm with a base of 10; L_1 and L_2 are the length and weight of the fish at the beginning and end of the period for which the growth ratio is calculated, respectively; and t_1 and t_2 are the time at the beginning and end of the period for which the growth ratio is to be calculated, respectively. Growth rates vary over a wide range; for example, for snapper in different waters, its growth coefficient varies between 0.97 and 7.22 eight times before maturity, but for mature snapper, it varies between 0.90 and 4.0 at four times.

When calculating growth rates or relative growth rates, the average length at each age, rather than the length of individual fish, is used in any context. Growth indicators can be used to classify the growth stages of fish in a given water. In the case of the offshore *L. polyactis*, its growth can be divided into three growth stages: the first stage is the vigorous growth stage, which occurs during the first 1 or 2 years of life, when a fish has not yet reached sexual maturity and body length growth is rapid. The second stage is the stable growth stage, from the second year to the sixth year; the gonads gradually mature; in the second year, there are still some fish gonads that are not fully mature or are close to mature, so the growth of this stage is stable. Growth in the second year can be as high as 53 mm. The third growth stage is senescence. Starting from the sixth year, growth slows, and a fish enters the senescence stage with the annual growth rate becoming very low (Table 4.2). In addition, changes in growth rates are influenced by a range of internal and external factors, such as relationships with water

Table 4.2 Growth status of *L. polyactis* in the northern East China Sea (Chen 2014; Chen and Liu 2017)

Age	Body length (mm)	Annual growth (mm) (L_2-L_1)	Growth rate % (L_2 – L_1)/L_1	Growth indicators (lgL_2 –lgL_1)/ 0.4343
1	139			
2	192	53	0.323	4.54
3	214	22	0.108	2.08
4	233	19	0.085	1.82
5	249	16	0.066	1.55
6	259	10	0.039	0.98
7	260	1	0.004	0.09
8	261	1		

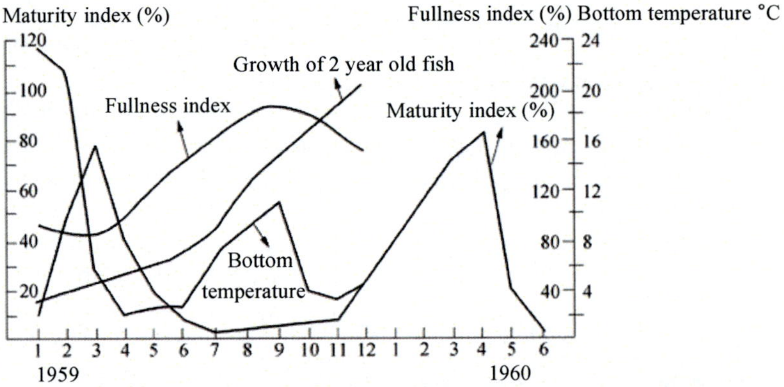

Fig. 4.18 Growth rate variation factors of *L. polyactis* in the southern Yellow Sea (Chen 2014; Chen and Liu 2017)

temperature, fish maturation systems, and fullness index by month (Fig. 4.18).

4.8.3.4 Relationship Between Fish Length and Body Weight

Fish growth is the process by which the size of an individual (e.g., body length, anal length, mantle length, shell length and carapace length, and body weight) increases over time. Growth has a major influence on stock dynamics, and studies of stock growth patterns and their associated influencing factors can provide relevant parameters for stock assessment and fisheries management.

There is a significant correlation between body length and anal length in fish, body length and carapace length in shrimp and crabs, mantle length in cephalopods, and shell length and body weight in shellfish (Figs. 4.19 and 4.20). This correlation is usually expressed by the following equation:

$W = a \times L^b$ $lnW = ln\ a + blnL$where W is body weight; L is body length, anal length, mantle length, shell length, and carapace length; a and b are two parameters to be determined; when $b = 3$, growth is allometry, and individuals have constant body shape and constant specific gravity; and when $b \neq 3$, growth is anisotropic. b values vary with fish growth and nutrition and between populations or between years in the same population.

The b values of marine fishes and invertebrates in China's coastal waters range from 2.4 to 3.2, with some differences between the same

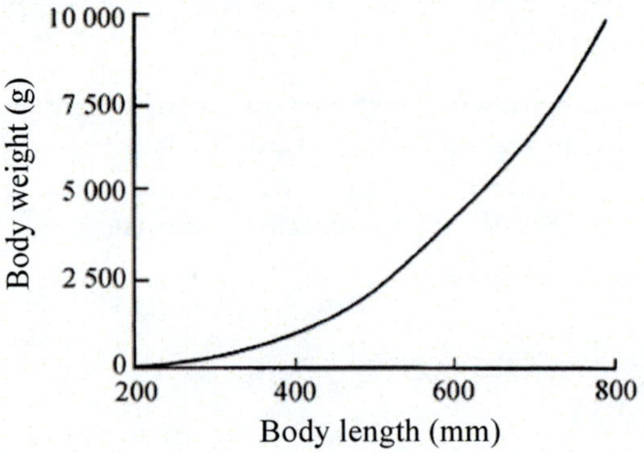

Fig. 4.19 Relationship between body length and body weight (Chen and Liu 2017)

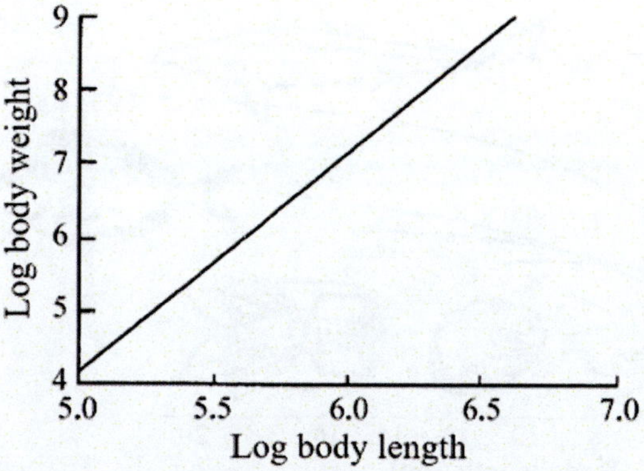

Fig. 4.20 Relationship between body length and body weight after taking the logarithms of both (Chen and Liu 2017)

population in different waters, life stages, and males and females. Freshwater fishes are slightly larger than marine fishes. Studies suggest that most fish have b values between 2.5 and 4.0. The biological significance is that when small changes in b values occur, changes in a values are more obvious; the magnitude of b values reflects changes in different populations, in the same population at different life stages, or in sex and the environment.

4.8.3.5 Fish Growth Equation

1) Fundamentals of growth. Von Bertalanffy viewed organisms as analogous to systems of chemical reactions in action, and according to the law of mass action, he attributed the physiological processes that determine the mass of an organism, at all times, to catabolism and synthesis. Based on general physiological concepts, Bertalanffy stated that the rate of anabolism is proportional to the rate of nutrient uptake, that is, to the size of the absorbing surface, while the rate of catabolism is proportional to the total biomass. With this concept in mind, the proposed Von Bertalanffy equation is the following:

$$\frac{dW}{dt} = HS - \beta W$$

where S is the effective physiological surface of the organism, H is the rate of substance synthesis per unit of "physiological surface," and β is the rate of substance decomposition per unit weight.

If an organism has homogeneous growth with constant specific gravity, then the above equation can be transformed by conversion to the following equation:

$$\frac{dW}{dt} = \alpha W^{2/3} - \beta W$$

In metabolic terms, growth is the instantaneous increase in body weight, proportional to body weight, and growth occurs as the difference between assimilation (i.e., substance synthesis) and anabolism. α represents the rate of assimilation, and β represents the rate of anabolism.

2) Von Bertalanffy's growth equation for body length and weight. Von Bertalanffy (1938) theoretically derived the equation for growth rate under the assumption that the weight of an organism is proportional to the cube of its length.

$$\frac{dL}{dt} = K(L_\infty - L)$$

Then, the equation can be solved.

$$L = L_\infty - C \times e^{-Kt}$$

Assuming t = t_0 and L = 0, we have $L_\infty - C \times e^{-Kt_0} = 0$ $C = L_\infty e^{Kt_0}$

and the following equations:

$$L_t = L_\infty \times \left[1 - e^{-K(t-t_0)}\right]$$

$$W_t = W_\infty \times \left[1 - e^{-K(t-t_0)}\right]^3$$

where L_t and W_t are individual length and weight at t, respectively; L_∞ and W_∞ are asymptotic length and weight, respectively; K is a growth parameter related to fish metabolism and growth; and t_0 is a hypothetical constant, i.e., the age at $W = 0$, which should theoretically be less than zero.

The entire growth process of an aquatic economic animal such as fish is represented in Fig. 4.21. The entire growth process shows an

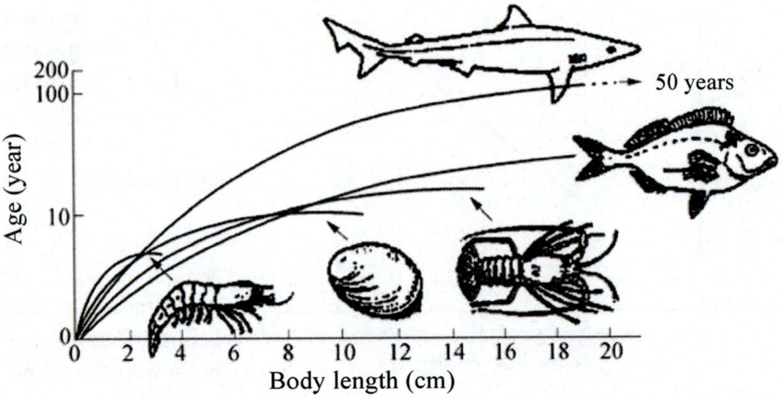

Fig. 4.21 Schematic diagram of the growth of fish and other aquatic economic animals (Chen 2014; Chen and Liu 2017)

asymptotic parabola, with body length reaching an asymptotic maximum.

4.9 Case Study: Age and Growth of Cephalopod Based on Statolith Microstructure

The Argentine shortfin squid *I. argentinus* is widely distributed in the waters of the Patagonian shelf in the southwestern Atlantic Ocean, which is rich in resources, and this species is an important fishery resource for coastal countries such as Uruguay, Argentina, and Brazil and plays an important role in the marine ecosystem. Cephalopod statoliths are widely used to study the age and growth of cephalopods and their life history due to their storage of rich information, corrosion resistance, and stable structure. Based on samples of *I. argentinus* collected by Chinese squid fishing vessel in the high seas of the southwestern Atlantic, statolith microstructural methods were used to study the age, growth, and population structure of *I. argentinus*.

4.9.1 Materials and Methods

4.9.1.1 Sample Collection

The samples were obtained by professionals on the squid fishing vessels. The samples were collected from February to May 2007, March to May 2008, and January to March 2010 in the high seas of the southwestern Atlantic Ocean. A total of 3462 samples were collected (308 samples in 2007, 262 samples in 2008, and 2892 samples in 2010), with 10–15 samples randomly selected from the catch at each station and the samples frozen and stored for transport to the laboratory.

4.9.1.2 Biological Measurements and Statolith Extraction

Biological measurements of *I. argentinus*, including mantle length (ML), body weight (BW), sex, and sexual maturity, were performed after thawing in the laboratory. Mantle length was determined to be 0.1 cm, and body weight was determined to be 0.1 g.

The statoliths were extracted from the statocyst, and 3450 pairs of intact statolith samples (2019 pairs of females and 1431 pairs of males) were obtained, with mantle lengths ranging from 267 to 350 mm and 122–266 mm for males and females of *I. argentinus*, respectively. The removed statoliths were numbered and stored in 1.5 ml centrifuge tubes containing a 95% ethanol solution to remove the soft membrane encasing the otoliths and the surface organic material.

4.9.1.3 Statolith Preparation and Age Determination

The statoliths were removed from the alcohol, the convex side (outer side) was placed in the prepared plastic mold, the prepared resin was poured in and covers the statolith, and after it hardened, the plane parallel to the long axis of

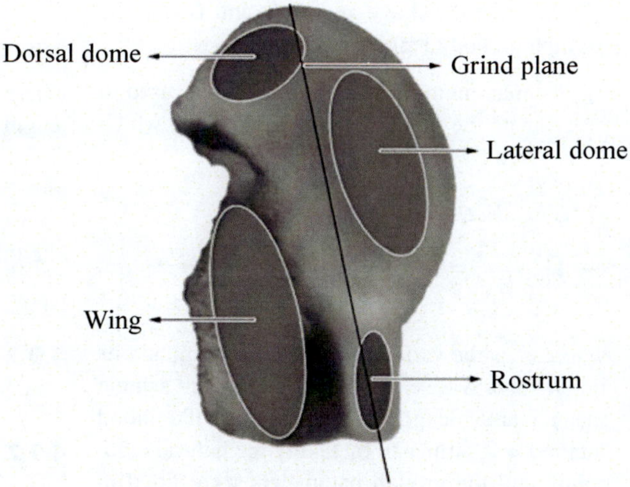

Fig. 4.22 Schematic diagram of the distribution and grinding planes of each zone of a statolith (Chen et al. 2014)

the statolith was chosen as the grinding plane for *I. argentinus* statolith (Fig. 4.22). Then, 2400 grits waterproof abrasive paper was used to grind from the wing area to the side area of the statolith, i.e., along the longitudinal plane to the core. Grinding was finished on one side, and then, the process was repeated on the other side. After both sides were ground to the core, the ground statolith slices were polished with a 0.05 μm alumina water flannel. Finally, the prepared statolith sections were placed in a scale bag for storage and marked.

The prepared statolith sections were removed and placed under an Olympus light microscope (objective ×4, ×10, and ×40, eyepiece ×10) ×400 and photographed using a CCD. The photographs were transferred to a computer via a data cable and then processed using Photoshop 8.0 image processing software, and the number of increments was counted. During the counting process, the increments of each statolith were counted twice, and the count was considered accurate if the difference between the number of increments and the mean value of each count was less than 5%; otherwise, two more counts were conducted to obtain the average of four counts. After random sampling and grinding of samples from different mantle length groups, 531 valid statoliths were finally obtained (160 samples in the year of 2007, 125 samples in the year of 2008, and 302 samples in the year of 2010)

4.9.1.4 Growth Model Selection

1) Analysis of covariance was used to determine whether there were significant differences in daily age and mantle length and age and body weight of *I. argentinus* between years and sexes.
2) Growth equations for *I. argentinus* were fitted using linear growth models, exponential growth models, power function growth models, logistic function models, and logistic, Von Bertalanffy, and Gompertz growth models.

$$\text{Linear equations}: L = a + bt$$

$$\text{Exponential equation}: L = ae^{bt}$$

$$\text{Power function equation}: L = aln(t) + b$$

$$\text{Logarithmic function equation}: L = aln(t) + b$$

$$\text{Logistic growth equation}: \quad L_t = \frac{L_\infty}{1 + \exp\left[-K(t_i - t_o)\right]}$$

$$\text{Von Bertalanffy}: L_t = L_\infty \times \{1 - \exp\left[-\mathrm{K}(t_i - t_o)\right]\}$$

$$\text{Gompertz}: \quad L_t = L_\infty \times \exp\{1 - \exp\left[-\mathrm{K}(t_i - t_o)\right]\}$$

where L is mantle length (or body weight) in mm or g; t is age in day; a, b, G, g, and k are constants;

t_O is the theoretical age at $L = 0$; and $L\infty$ is the asymptotic body length.

3) The maximum likelihood method was used to estimate the model growth parameters, which is given by the following equation:

$$L\left(\widetilde{L}|L_\infty, K, t_o, \sigma^2\right) = \Pi_{i=1}^{N} \frac{1}{\sigma\sqrt{2\pi}} \exp\left\{\frac{-[L_i - f(L_\infty, K, t_o, t_i)]^2}{2\sigma^2}\right\}$$

where σ^2 is the variance of the error term, and its initial value was set at 15% of the overall sample mean mantle length. The maximum likelihood method was estimated by taking the natural logarithm, and the growth parameters were fitted in Excel 2003 using the planning solution.

4) Akaike's information criterion (AIC) was applied to grow the model comparison. It was calculated using the following equation:

$$\text{AIC} = -2\ln\text{L}\left(p_1, \ldots, \text{p}_m, \sigma^2\right) + 2m$$

where L (p_1,..., p_m,) is the maximum likelihood value of the mantle length at the daily age, δ^2 is the maximum likelihood estimate of the model parameters, and m is the number of parameters to be estimated in the model. Among the seven growth models, the model that achieved the smallest AIC value was the most appropriate growth model.

4.9.1.5 Growth Rate Estimation

The instantaneous relative growth rate G and the absolute growth rate AGR were used to analyze the growth of *I. argentinus* with the following equations:

$$G = \frac{ln\,(R_2) - ln\,(R_1)}{t_2 - t_1} \times 100$$

where R_2 is the body weight (BW) or mantle length (ML) at age t_2, R_1 is the body weight (BW) or mantle length (ML) at age t_1, and G is the relative growth rate.

$$AGR = \frac{R_2 - R_1}{t_2 - t_1}$$

where R_2 is the body weight (BW) or mantle length (ML) at age t_2, R_1 is the body weight (BW) or mantle length (ML) at age t_1, ML is in mm, BW is in g, and AGR is in mm $^{d-1}$ or g $^{d-1}$.

A time interval of 30 days was used to calculate the growth rate.

4.9.2 Individual Composition of the *I. argentinus*

4.9.2.1 Compositions of Mantle Length and Weight

In 2007, the average mantle length of the samples ranged from 178 to 346 mm, with the samples measuring 231.72 mm, 266.88 mm, 266.88 mm, 231.72 mm, and 266.88 mm, and the dominant carcass length group was 180–270 mm, accounting for 87.95% of the total amount of carcasses measured. In 2008, the mantle length of the sample ranged from 193 to 364 mm, with a mean mantle length of 266.88207 mm, and the dominant mantle length group was 210–330 mm, accounting for 89.35% of the total (Fig. 4.23a).

In 2007, the weight range of the samples was 102–802 g with a mean weight of 440.86 g, and the dominant recombination was 100–350 g (350 g, 300 g, 300 g, and 700 g), accounting for 82.84% of the total. In 2010, the weight range was 187.11 g to 425 g, and the dominant recombination was 100–300 g, accounting for 91.51% of the total (Fig. 4.23b).

4.9.2.2 Age Composition

The statolith microstructure image showed (Fig. 4.24) that the age range of the samples in 2007 was 207–370 d, with a mean age of 286.5 d, and the dominant age group was 240–330 d, accounting for 83.19% of the total samples; the age range of the samples in 2008 was 208–359 d, with a mean age of 293.8 d, and the dominant age group was 240–330 d, accounting for 91.72% of the total; the age range of the samples in 2010 was 173–400 d, with a mean age of 300 d, and the dominant age group was 240–360 d, accounting

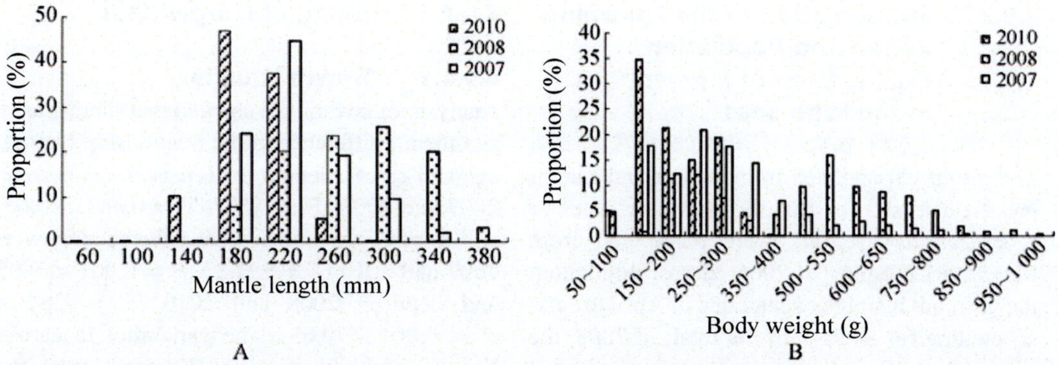

Fig. 4.23 Distribution of mantle length (**a**) and weight (**b**) of *I. argentinus* in different years (Chen et al. 2014)

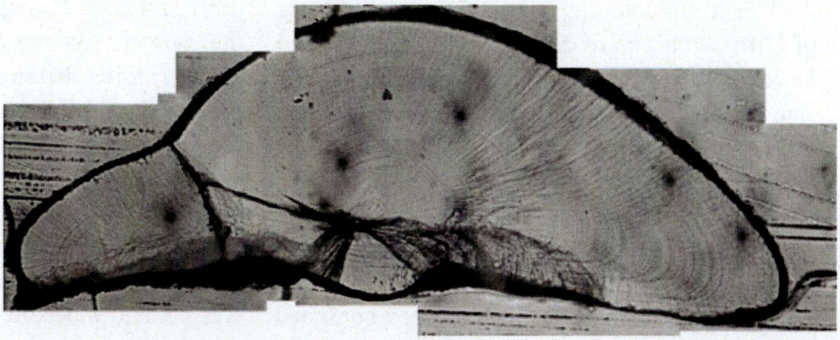

Fig. 4.24 Schematic microstructure superimposed on statoliths of *I. argentinus* after grinding (female, carcass length 255 mm, body weight 255 g, age 315 d) (Chen et al. 2014)

for 88.68% of the total (Fig. 4.25). In 2008, the age range of the samples was 208–359 d with a mean age of 293.8 d, and the dominant age group was 240–330 d, accounting for 91.72% of the total; in 2010, the age range of the samples was 173–400 d with a mean age of 300 d, and the dominant age group was 240–360 d, accounting for 88.68% of the total (Fig. 4.25).

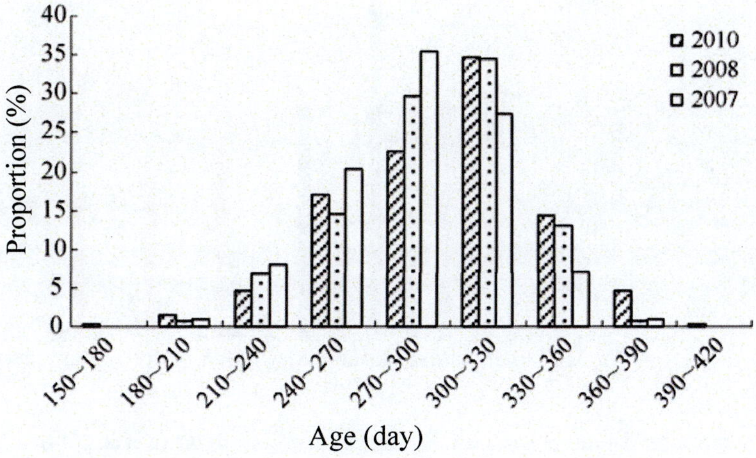

Fig 4.25 Age distribution of *I. argentinus* across years (Chen et al. 2014)

4.9.3 Extrapolation of the Spawning Period and Population Delimitation of *I. argentinus* in the High Seas

The results extrapolated from the age and date of harvest data showed that the spawning dates of *I. argentinus* in 2007 were distributed from March to December in 2006, almost throughout the year, but mainly concentrated in April to July, accounting for 84.91% of the total; in 2008, the spawning dates of *I. argentinus* were distributed from May to December in 2007, mainly concentrated in June to August, accounting for 90.94% of the total. In 2010, the spawning dates of *I. argentinus* were distributed from January to December 2009 (Fig. 4.26), but the peak spawning period occurred from March to May, accounting for 74.77% of the total, followed by from June to August, accounting for 13.59% of the total.

Based on the extrapolation of the incubation period data, it can be assumed that the *I. argentinus* catches in the high seas consisted of two main spawning groups, autumn and winter, with the 2007 and 2008 *I. argentinus* being mainly the winter spawning group (June–July) and the 2010 sample being mainly the autumn spawning group (March–May).

4.9.4 Growth of *I. argentinus*

4.9.4.1 Growth Equation

Analysis of covariance showed that there was no significant difference in the relationship between age and carcass length for *I. argentinus* between 2007 and 2008 ($F = 0.597$, $P = 0.082 > 0.05$), while there was a significant difference between 2007 and 2010 ($F = 227.33$, $P = 0.001 < 0.05$) and between 2008 and 2010 ($F = 264.44$, $P = 0.001 < 0.05$). The variability in carcass length growth that existed between years was also explained by the variability in carcass length growth that existed between the groups, as the 2007 and 2008 *I. argentinus* samples belonged to the winter spawning group, while the 2010 *I. argentinus* belonged to the fall spawning group. Based on the analysis of covariance, there were sex differences in carcass length growth between the winter cohort ($F = 161.36$, $P = 0.003 < 0.05$) and the fall cohort ($F = 65.56$, $P = 0.001 < 0.05$); therefore, the 2007 and 2008 samples were combined, while the 2010 sample was studied independently and by sex. The relationship between age and carcass length of *I. argentinus* was investigated by fitting the equations and the maximum and minimum mantle lengths. By fitting the equations, optimizing the maximum likelihood rule, and comparing the

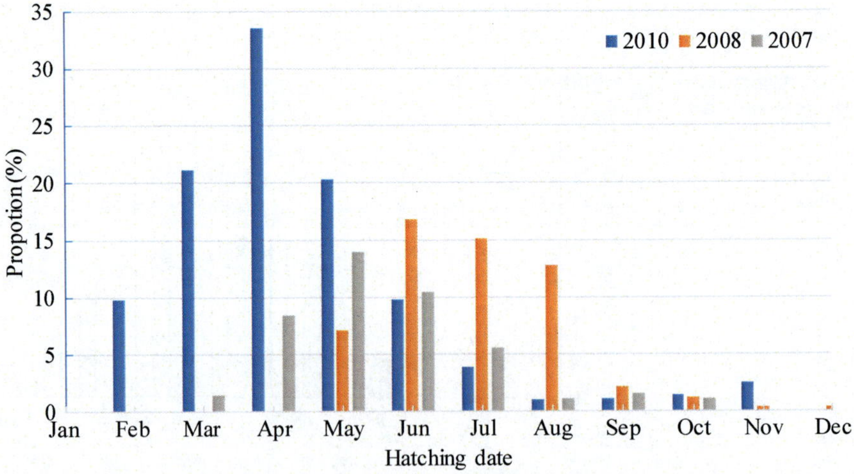

Fig. 4.26 Distribution of spawning dates for *I. argentinus* (Chen et al. 2014)

AIC, the best growth equations for the mantle length of *I. argentinus* were obtained as follows:

Winter spawning aggregations:

Females: $ML = 106.9955 \times e^{0.0032Age}$ ($R^2 = 0.7082$, n = 152)

Male: $ML = 0.6705 \times Age + 43.101$ ($R^2 = 0.5756$, n = 109)

Fall spawning aggregations:

Female: $ML = 116.65 \times e^{0.0021Age}$ ($R^2 = 0.5582$, n = 141)

Male: $ML = 129.6903 \times \ln(Age) - 531.0295$ ($R^2 = 0.6478$, n = 127)

4.9.4.2 Relationship Between Age and Weight

Analysis of covariance showed that there was no significant difference in the relationship between the age and body weight of *I. argentinus* between 2007 and 2008 ($F = 13.3274$, $P = 0.08 > 0.05$), while there was a significant difference between 2007 and 2010 ($F = 220.64$, $P = 0.001 < 0.05$) and between 2008 and 2010 ($F = 515.26$, $P = 0.001 < 0.05$). Analysis of covariance revealed sex differences between the winter spawning group ($F = 70.54$, $P = 0.003 < 0.05$) and the fall spawning group ($F = 1.748$, $P = 0.001 < 0.05$); thus, the 2007 and 2008 samples were combined, while the 2010 sample was separated into sexes to study the relationship between age and weight of *I. argentinus*. The relationship between age and body weight was investigated by fitting the maximum and maximum weight equations. By fitting the equations, optimizing the maximum likelihood rule, and comparing the AIC, the following growth equations were obtained for the body weight of *I. argentinus*:

Winter spawning aggregations:

Females: $BW = 9.09 \times 0^{-5Age2.7062}$ ($R^2 = 0.6959$, n = 152)

Males: $BW = 15.5689 \times e^{0.0101Age}$ ($R^2 = 0.6319$, n = 109)

Fall spawning aggregations:

Females: $BW = 34.3861 \times e^{0.0061Age}$ ($R^2 = 0.5413$, n = 141)

Male: $BW = 404.3661 \times \ln(Age) - 2101.6$ ($R^2 = 0.5407$, n = 127)

4.9.4.3 Growth Rate Analysis

Studies have shown that *I. argentinus* grow relatively rapidly. For the winter spawning population, the mean relative and absolute growth rates for female mantle length were 0.29% d^{-1} and 0.73 to 1.09 mm d^{-1}, respectively, with maximum relative (0.39% d^{-1}) and absolute (1.09 mm d^{-1}) growth rates occurring from 300 to 330 d; minimum relative (0.14% d^{-1}) and absolute (0.37–1.09 mm.d^{-1}) growth rates occurred from 270 to 300 d; mean relative and absolute growth rates for male mantle length were 0.20% d^{-1} and 0.50 mm.d^{-1}, respectively, with maximum relative (0.35% d^{-1}) and absolute (0.84 mm.d^{-1}) growth rates occurring from 300 to 310 d; the minimum relative growth rate (0.04% d^{-1}) and absolute growth rate (0.09 mm.d^{-1}) occurred from 210 to 240 d.

The mean relative and absolute growth rates of female mantle length in the fall spawning aggregation were 0.19% d^{-1} and 0.41 mm d^{-1}, respectively, with the maximum relative (0.29% d^{-1}) and absolute (0.65 mm d^{-1}) growth rates occurring from 300 to 330 d. The minimum relative (0.07% d^{-1}) and absolute (0.15 to 0.65 mm d^{-1}) growth rates occurred from 240 to 270 d; the mean relative and absolute growth rates of male mantle length were 0.19% d^{-1} and 0.38 mm d^{-1}, respectively, with the maximum relative (0.70%) growth rate and absolute growth rate (0.38 mm d^{-1}) occurring from 240 to 270 d; the minimum relative growth rate (0.05% d^{-1}) and absolute growth rate (0.12 mm d^{-1}) occurred from 390 to 420 d.

Overall, for both the winter spawning and fall spawning groups, both relative and absolute growth rates were greater for the female samples than for the male samples within the same age range, and there was an overall decreasing trend in relative growth rates and an overall increasing trend in absolute growth rates for mantle length for both groups as age increased (Fig. 4.27).

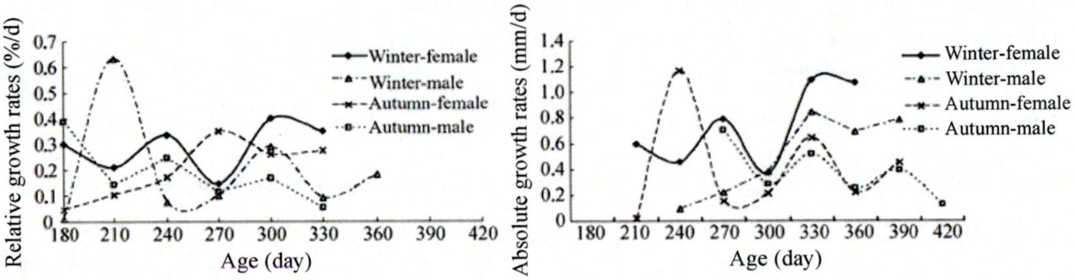

Fig. 4.27 Mantle length growth rate of *I. argentinus* (Chen et al. 2014)

The mean relative and absolute growth rates by body weight of females in winter spawning aggregations were 0.91% d^{-1} and 3.25 g d^{-1}, respectively, with the maximum relative growth rate (1.28% d^{-1}) and absolute growth rate (6.35 g d^{-1}) occurring from 300 to 330 d and 330 to 360 d, respectively; the minimum relative growth rate (0.56% d^{-1}) and absolute growth rate (1.08 g d^{-1}) occurred from 210 to 240 d; the mean relative and absolute growth rates of males in the winter spawning aggregations were 0.89% d^{-1} and 3.08 g d^{-1}, respectively; the average relative and absolute growth rates of male body weight in the winter spawning aggregations were 0.89% d^{-1} and 3.35 g d^{-1}, respectively, with the maximum relative growth rate (1.81% d^{-1}) and absolute growth rate (5.94 g d^{-1}) occurred from 300 to 300 d and 360 to 390 d, respectively; the minimum relative growth rate (0.69% d^{-1}) and absolute growth rate (1.08 g d^{-1}) occurred from 210 to 240 d.

Fall female body weights had averaged relative and absolute growth rates of 0.50% d^{-1} and 1.04 g d^{-1}, respectively, with maximum relative (0.56% d^{-1}) and absolute (1.62 g d^{--1}) growth rates occurred from 330 to 360 d; minimum relative (0.34% d^{-1}) and absolute (0.45 g d^{-1}) growth rates occurred from 210 to 240 d; mean relative and absolute growth rates for male body weight were 0.59% d^{-1} and 1.45 g d^{-1}, respectively, with maximum relative (0.89% d^{-1}) and absolute (1.91 g d^{-1}) growth rates both occurred from 300 to 300 d; minimum relative (0.06% d^{-1}) and absolute growth rates (0.18 g d^{-1}) occurred from 390 to 420 d.

Overall, for both the winter spawning and fall spawning groups, the female samples were generally larger than the male samples within the same age range for both relative and absolute growth rates; the relative growth rates for both groups showed a decreasing trend in body weight with increasing age, while the absolute growth rates showed an overall increasing trend (Fig. 4.28).

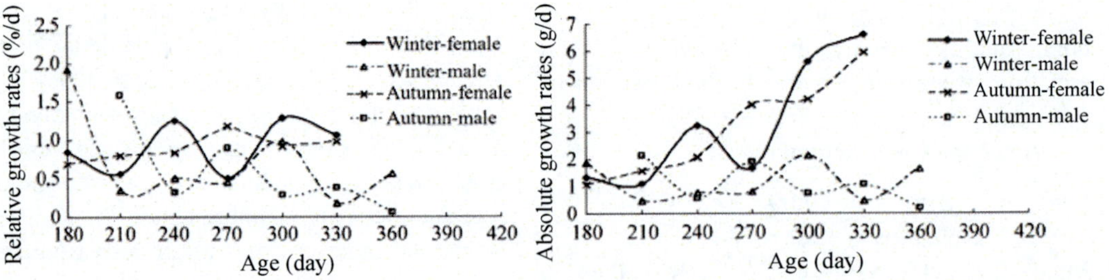

Fig. 4.28 Distribution of body weight growth rate of *I. argentinus* (Chen et al. 2014)

References

Campana SE, Zwanenburg CT, Smith JN (1990) $^{210}Pb/^{226}Ra$ Determination of longevity in redfish. Can J Fish Aquat Sci 47(1):162–165

Chen XJ (2014) Fisheries resources and fisheries oceanography. Ocean Press (In Chinese)

Chen XJ, Liu BL (2017) Fishery resources biology. Science Press (In Chinese)

Chen XJ, Lu HJ, Fang Z et al (2014) Fishery biology of *Illex argentinus* in the Southwest Atlantic Ocean. Science Press

Clarke MR (1962) The identification of cephalopod beaks and the relationship between beak size and total body weight. Bull Br Mus Nat Hist 8:419–480

Clarke MR (1965) "Growth rings" in the beaks of the squid *Moroteuthis ingens* (Oegopsida: Onychoteuthidae). Malacologia 3(2):287–307

Ettershank G (1984) A new approach to the assessment of longevity in the Antarctic krill *Euphausia superba*. J Crust Biol 4(1):295–230

Fan RQ, Jiang M, Ru SG et al (2000) The scanning electron microscope observation on the surface structure of scales of *Oreochromis aureus*. Mar Sci 3:48–51. (In Chinese)

Fridriksson A (1934) On the calculation of age-distribution within a stock of cod by means of relatively few age determinations as a key to measurements on a large scale. Rapp P V Reun Cons Int Explor Mer 86:1–14

Hernández-López JL, Castro-Hernández JJ, Hernández-García V (2001) Age determined from the daily deposition of concentric rings on common octopus (*Octopus vulgaris*) beaks. Fish Bull 99:679–684

Hu GY (2016) Microstructure and microchemistry of beak for jumbo flying squid *Dosidicus gigas* off the Peruvian Exclusive Economic Zone. Shanghai Ocean University (In Chinese)

Hueck W (1912) Pigmentstudien. Beitrage zur pathologischen Anatomie und zur allgemeinen Pathologie 54:68–232

Isermann DA, Knight CT (2005) A computer program for age–length keys incorporating age assignment to individual fish. N Am J Fish Manag 25:1153–1160

Kastelle CR, Kimura DK, Jay SR (2000) Using $^{210}Pb/^{226}Ra$ disequilibrium to validate conventional ages in Scorpaenids (genera *Sebastes* and *Sebastolobus*). Fish Res 46(1–3):299–312

Li CH, Sha XS, You F et al (1993) Daily growth increments in otoliths of mullet larvae, *Mugil So-Iuy* BASILEWSKY and determination from filed-collected ones. Mar Lakes 4:345–349+447 (In Chinese)

Liu BL, Chen XJ, Fang Z, et al (2017) Cephalopod beaks. Science Press (In Chinese)

Nixon M (1973) Beak and radula growth in Octopus vulgaris [J]. J Zool 170:451–462

Ogle DK, Isermann DA (2017) Estimating age at a specified length from the von Bertalanffy growth function. N Am J Fish Manag 37:1176–1180

Raya CP, Hernández-López CL (1998) Growth lines within the beak microstructure of the Octopus vulgaris Cuvier, 1797. S Afr J Mar Sci 20:135–142

Sheehy MRJ (1990) Individual variation in and the effect of rearing temperature and body size on the concentration of fluorescent morphological lipofuscin in the brain of freshwater crayfish, *Cherax cuspidatus* (Crustacea: Parastacidae). Comp Biochem Physiol 96A:281–286

Smale MJ, Clarke MR, Klages NTW, Roeleveld MAC (1993) Octopod beak identification – resolution at a regional level (Cephalopoda, Octopoda: South Africa). S Afr J Mar Sci 13:269–293

Von Bertalanffy L (1938) A quantitative theory of organic growth (inquiries on growth laws II). Hum Biol 10:181–213

Zhu GP, Song Q (2016) Aging crustacean based on the lipofuscin analysis: a review. J Ecol 8: 2225–2233 (In Chinese)

Sexual Maturation, Reproductive Habits, and Fecundity of Fish

Xinjun Chen, Bilin Liu, and Dongming Lin

Abstract

The reproduction activity of fishery resource population is one of the most important components of its life activities, and is the most important activity of multiplying population and preserving species. The unique reproduction characteristic of each fishery resource stock is one of the adaptive attributes of the population to the living conditions in waters. The research on the mechanism and strategy of fish reproduction can not only provide important scientific basis for rational fishing and scientific formulation of fishery management measures but also have great practical significance for the correct solution of artificial reproduction and artificial release. Reproduction research of fishery resources stocks is involved in developmental biology, genetics, physiology, and ecology. Fish reach sexual maturity characteristics, spawning population structure, and its relationship with environmental factors, fertility and reproductive strategy has become a routine of fisheries resources investigation and research. Therefore, the key content of this chapter is to understand the process of fish sexual maturation, learn the method of gonad maturity division, master the basic method of breeding habits and fertility measurement, and understand the fish reproductive strategy and its relationship with the environment. Cephalopods are important marine economic animals and play an indispensable role in marine ecosystem; therefore, in this chapter, cephalopods' growth and maturation, fertility characteristics, and reproductive strategy and the nature of their eggs and oocyte development are introduced.

Keywords

Sexual maturation · Reproductive habits · Fecundity of fish · Cephalopods

Abbreviations

C_x	catch of females likely to spawn in an age group
E_i	the total ovary egg carrying capacity
E_p	population fecundity
FE	the absolute fecundity
fe	the number of eggs in the sample
F_x	the average individual fecundity (number of eggs) for the same age group
ICES	International Council for the Exploration of the Sea
$J_{NG,L}$	Ovary length index
$J_{NL,s}$	Ovary body index
$J_{T,L}$	Spermatozoa length index

X. Chen (✉) · B. Liu · D. Lin
College of marine sciences, Shanghai Ocean University, Shanghai, China
e-mail: xjchen@shou.edu.cn; bl-liu@shou.edu.cn; dmlin@shou.edu.cn

X. Chen, B. Liu (eds.), *Biology of Fishery Resources*, https://doi.org/10.1007/978-981-16-6948-4_5

$J_{T,s}$	Spermatozoa body index
N_x	the number of females likely to spawn in an age group (number of tails)
RS	the remaining spawning population
SP	the spawning population
SS	the supplementary spawning population
U	the weight of the ovary part
V	the volume of the ovary
W_s	the weight of the sample
W_w	the weight of the whole ovary

5.1 Sexual Characteristics of Fish and Their Sexual Maturation

5.1.1 Difference Between Male and Female Fish

Many female and male fishery resource species cannot be distinguished by their external genitalia, as is the case with mammals. In many fish, sexes are often distinguished only by other parts of the fish or by features such as body size, body color, and the outer opening of the reproductive orifice. In some fish, the difference between the sexes, such as pearl stars and marital coloration, is apparent during the breeding season and is often particularly pronounced in females. This type of feature is often referred to as a secondary sexual characteristic or hermaphroditism. However, there are also a number of fishes in which males are not easily distinguished from males by their appearance.

The sexual characteristics of fish are divided into three types: (1) dioecious, (2) normal hermaphrodites, and (3) individuals who are partly dioecious and partly hermaphroditic or sexually reversed within the same species. Thus, the distinction between the sexes in fish is not as simple as one might think. Scientific distinctions between the sexes in fish sometimes even require histological observations.

The external characteristics of the fish, such as body shape, fins, reproductive pores, and olfactory bulbs, are generally used to distinguish between the sexes. Male differences in morphological characteristics are a common method of determining male and female fish. The differences in morphological characteristics between the sexes are generally stable, but some fish undergo significant changes in body shape during the breeding season. For example, during the anadromous spawning migration of *Oncorhynchus keta*, both jaws of the male fish are bent into a hooked shape, and the males develop giant teeth; the back of the male fish of *O. gorbuscha* has a distinct bulge, so it is also called humpback cannabis salmon.

In some fish, the structures of the reproductive pores differ between males and females; e.g., *Oreochromis* females have a shorter reproductive papilla and reproductive pore followed by a urinary pore after the anus, while males have only a longer cloacal papilla after the anus and a combined reproductive and urinary pore. This scenario also occurs in *Pagrus major*.

The size of the olfactory bulb and the number of olfactory plates can be used to distinguish male from female *Lophiodes* spp.; male *Lophiodes* spp. have olfactory bulbs that are two to three times larger and olfactory plates that are one time larger than those of the female *Lophiodes* spp. that are of the same length. The external reproductive organs of the fish enable identification of the two sexes. It should be noted, however, that some fish do not have external reproductive organs.

Many fish appear brightly colored during the breeding season, or their existing coloration becomes more vivid, which is generally more prominent in males and disappears after the reproductive season; this coloration is called marital coloration. For example, *O. keta* have a silvery body when living in the sea and become dark brown when they are anadromous and migrate during the breeding season, while males also have bright red spots on their sides.

During the breeding season, individual body parts of some fish (e.g., operculum, fin rays, muzzle, and back of head) develop hard white vertebral projections called pearl stars, which are the result of extraordinarily hypertrophic and keratinized epidermal cells. Most pearl stars are most frequently found only in males, but in some species, they are present in both sexes during reproduction. This feature is more common in the carp family; for example, the four major carp species, *Mylopharyngodon piceus*, *Ctenopharyngodon idellus*, *Hypophthalmichthys molitrix*, *Hypophthalmichthys nobilis*, all have pearl stars on their pectoral fin rays. It is generally believed that pearl stars excite and stimulate both male and female fish during spawning and sperm production, and when spawning occurs, male and female body parts which are in contact with each other are mostly where the pearl stars are dense.

The vast majority of fish species are dioecious, but some species, such as a few in Sparidae, are sometimes hermaphroditic with both ovarian and spermatogonial tissues in the gonads. A typical example of a fish that experiences sexual reversal, or gonadal transformation, is *Monopterus albus*. Although this species has both sexes, it is female from the embryonic stage to sexual maturity, and only after maturation and spawning do some individuals change to males.

In summary, in fish, the distinction between sexes is generally based on external characteristics, but to accurately determine sex, dissection is often required to observe the internal reproductive systems of fish and to distinguish between male and female fish.

5.1.2 Sexual Maturation Process and Biological Minimum Size

5.1.2.1 Sexual Maturation

The timing of the onset of sexual maturity in fish is dependent on the species and is an adaptation that has developed over time in fish under different environmental conditions. Usually, the time of sexual maturation has a large range of variation and varies within a population. Studying the sexual maturation process of fish is important for estimating population change trends and the appropriate exploitation and utilization of fishery resources.

Within the same population, early maturation of fish gonads is related primarily to individual size (usually body length). It has been suggested that sexual maturation begins when fish reach approximately half their maximum length, so the faster a fish grows, the earlier it matures sexually.

During the maturation process of fish of the same age, those that have reached or will reach sexual maturity are larger in length than those that are sexually immature. Regardless of age, the maturation ratio of fish of all ages increases monthly as fish grow. Although the timing of birth for striped bass varies, the size of striped bass that reach sexual maturity for the first time is approximately the same, at approximately 180 mm (Fig. 5.1). Thus, the sexual maturity of *T. lepturus* is more closely related to length than to age.

In general, there is considerable interspecific variability in the age of first sexual maturity for continuously spawning fish species. Generally, populations that sexually mature earlier have a short life cycle and rapid generational renewal; populations that sexually mature later have a long life cycle and slow generational renewal. In the tropics and subtropics, Cyprinodontidae reach sexual maturity in just a few weeks; Acipenseridae, such as *Huso dauricus*, do not begin to reach sexual maturity until they are 17 years old; and the age of sexual maturity in Pleuronectiformes varies from 1 to 15 years. In Chinese marine fishes, the age of sexual maturity is usually 1–2 years. It is worth noting that an early sexual maturity age within the same stock is a sign of early development due to the decline in resources caused by overfishing or warming water temperatures, and this early maturation is also an adaptive characteristic of the stock to reproduce and rear offspring.

5.1.2.2 Size at Maturity in Fish

After eggs are fertilized and hatch, juvenile fish gradually grow, and after a certain level of growth, their gonads begin to develop and mature. The time when sexual maturity begins varies from fish to fish; that is, the same species of fish, depending on where they live, may begin to mature sexually at different times. This time when gonads begin to develop and mature after a certain level of growth in a juvenile fish is generally referred to as the time of first sexual maturity. The minimum length at which a fish reaches its initial gonadal time is called the biological minimum, and this body length is usually at the

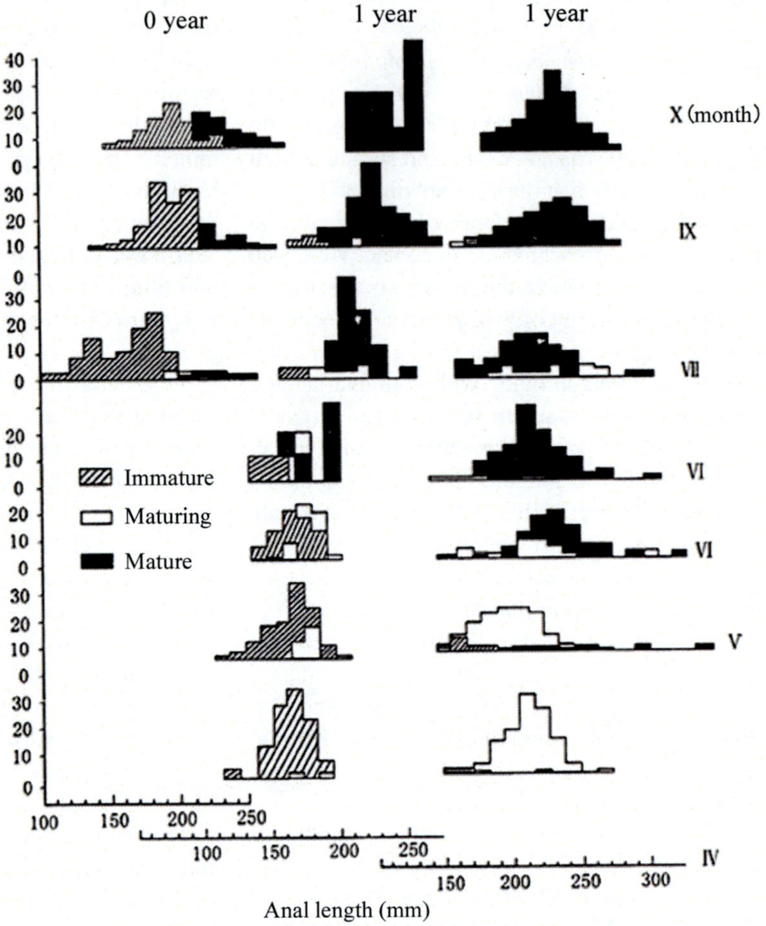

Fig. 5.1 Relationship between sexual maturity and growth in *T. lepturus* (Chen 2014; Chen and Liu 2017)

inflection point of the von Bertalanffy weight growth equation.

The time to first sexual maturity is related to when a fish reaches a certain body length and is less related to the amount of time a fish has lived. The faster the fish grow, the shorter the amount of time it takes for that fish to reach sexual maturity and vice versa. For most fish, those living in high latitude waters usually begin to sexually mature later than those living in low latitude waters, and the timing of sexual maturation differs between males and females. For example, within the same *P. crocea* group, sexual maturity in the fish living in the coastal area of Zhejiang Province starts at the age of 2; sexual maturity in a large number of male and female fish begins at the ages of 3 and 4, respectively; and both male and female fish are sexually mature by the time they reach the age of 5. For the fish living off Scammon Island in the eastern part of Hainan Island, a few individuals start to mature sexually at the age of 1, and a large number mature at the ages of 2 and 3. In the Northern Hemisphere, the age that *P. crocea* reach first sexual maturity gradually occurs early from north to south, males sexually matures earlier than females, and the body length and weight of males at sexual maturity were smaller than those of females. For example, many male greater amberjack off Zhejiang were 250 mm long and weighed approximately 200 g at the start of sexual maturation, while most females were 280 mm long and weighed approximately 300 g at the start of sexual maturation (Chen and Liu 2017).

In terms of fishing activities, on the one hand, it is necessary to consider the biological characteristics of the fishing objects, understand their biological minimum sizes, and set minimum catchable standards to ensure a certain number of adult fish so that the fishing stock has sufficient replenishment, and on the other hand, it is important to achieve the maximum utilization of the resources, that is, to obtain the maximum biomass. Therefore, the correct determination of the biological minimum size of a fish is of great significance to the conservation of fishery resources when setting the catch criteria.

5.1.2.3 Sexual Maturity and the External Environment

Throughout the life history of fish, they are influenced by environmental factors. During sexual maturation and spawning, external environmental factors (both biological and nonbiological) are more significantly related to fish because of their more stringent requirements.

During spawning, some fish have very strict requirements related to environmental factors or a certain factor; for example, anadromous fish must migrate through difficult conditions to a specific river to spawn each year; fish such as *P. crocea* cannot spawn without a certain water flow level, even if the temperature and dissolved oxygen are suitable.

Since most fish need to grow to a certain size to mature and spawn and food availability is the most important condition related to growth, there is also a close relationship between prey conditions and how early or late fish spawn.

Generally, fishery resources mature earlier in water bodies with high average temperature, long light hours, abundant prey, and excellent water quality conditions (e.g., dissolved oxygen and pH). For example, fish in the South China Sea mature earlier than those in the Yellow Sea and East China Sea. Because fish spawning has a close relationship with environmental factors, this relationship can be used in fishery forecasting and analysis.

5.1.2.4 Sex Ratio

The sex ratio is the ratio of the number of females to males in a fish population and can usually be expressed as the ratio of the number of females to males in a catch.

A proper female to male ratio is important for reproductive effectiveness. Maintaining a certain female to male ratio will result in continuous reproduction and offspring survival. Thus, the female to male ratio is a specific expression of biological characteristics.

The sex ratios of fish vary depending on the habitat and season. Take, for example, *Muraenesox cinereus*; in the East China Sea, there are more females than males in winter, more males than females in spring, and similar sex ratios in summer and autumn; however, in the Kyushu region of Japan, sex ratios also vary considerably in winter and spring but are the opposite of those in the East China Sea.

Sex ratios also vary with the size of individuals in a population. For example, in the East China Sea, female *Dentex tumifrons* account for 70% of the population at the smaller stage, 50% at 220 mm, and only 10–20% at the older stage. In another example, when the average length of *D. tumifrons* is less than 280 mm, the majority is male; at a length of 360 mm, the sex ratio is equal to 1:1; at a length of more than 360 mm, the majority of the population is female.

The sex ratio of fish also varies with life stage; e.g., in the Eastern Sea, there are more female *L. polyactis* than male in February–March and May–August, and the sex ratio of male to female is similar from October to January. The average sex ratio at fish spawning grounds is 1:1.

The overall sex ratio of fish during reproduction is generally close to 1:1 but varies slightly during the various stages of reproduction. In the early stages of reproduction, males generally predominate, the sex ratio is approximately equal during the reproductive period, and the proportion of males gradually increases in the later stages of reproduction. In a spawning population, more males are small, and more females are large. This scenario occurs because males are sexually mature and therefore participate in the spawning population earlier than females but generally have a shorter life span, so there are fewer larger males (older fish), which is important for the prosperity of the population, as the early mortality of males ensures that offspring and females receive plenty of prey.

However, during the reproductive season, the sex ratios of spawning aggregations at spawning sites often differ significantly, sometimes with more males than females. For example, an analysis of the sex ratios of the five main reproductive groups of *P. crocea* (Table 5.1) showed that all sizes of males outnumbered females in a relationship of approximately 2:1. It has been suggested that such sex ratios are relevant for ensuring egg fertilization rates and increasing the number of offspring as an adaptation to reproductive conditions, as the greater amberjack spawns in very fast-flowing waters.

In conclusion, the diverse fish sex ratios are the result of the adaptations of different fish to the diversity of their living environments. Thus, sex ratios are greatly important to the study of fishery resources biology.

5.1.3 Methodologies for Studying Gonadal Maturity

5.1.3.1 Visual Grade Method

Determining fish gonadal maturity is one of the most common actions in fishery resource survey studies, and the most common and practical method is the visual method, in addition to histological methods. In practice, the results observed by visual inspection can essentially meet the needs of determining maturity. The criteria for classifying maturity levels by the visual method are mainly based on characteristics such as the shape of the gonads, the distribution of blood vessels, and the condition of eggs and semen. The criteria used by European and American countries, the former Soviet Union, and Japan are not exactly the same; for example, European and American scholars usually use the improved criteria for determining the maturity of Atlantic herring gonads proposed by Hjort in 1910, which was adopted by the International Council for the Exploration of the Sea (ICES), and these criteria classify fish sexual maturity into seven grades. Soviet scholars, on the other hand, often use the six-stage division, and Japanese scholars divide fish gonadal maturity into five stages, namely, the resting, immature, mature, matured, and late-spawning stages.

The gonadal maturity criteria used in China are generally those proposed by K.A. Kiselevich in 1922. These criteria have been applied for a long time with good practical results and have been slightly modified and compiled in the Manual of Marine Fisheries Resources Transfer (Yellow Sea Fisheries Research Institute 1981). However, regardless of what criteria are used to classify fish maturity, the following requirements should be considered: (1) maturity classes must correctly reflect the variation in the development of fish gonads; (2) maturity classes should be developed in accordance with the biological characteristics of the fish; (3) to determine the division of stages, the variation in external characteristics visible to the naked eye and internal characteristics invisible to the naked eye must be estimated in the classes; and (4) the division grades should not be excessive so that they can accommodate field work.

The six stages of gonadal maturity commonly used in China to classify fish are described as follows:

Table 5.1 Sex ratios of the five major reproductive groups of *P. crocea*

Reproductive group	Lvsi	Daiqu	Maotou	Guanjing	Naozhou
Male (%)	66.0	72.82	81.86	69.11	69.96
Female (%)	34.0	27.18	18.14	30.89	30.04
n	2370	5814	1803	2289	4434

Chen and Liu (2017)

Stage I: Sexually immature individuals. Gonads are undeveloped and attached immediately to the inner side of the body wall in a thin line or band. Males and females cannot be identified by the naked eye.

Stage II: Individuals whose gonads have begun to develop or redevelop after spawning. The finely banded gonads have thickened and can be identified as male and female. The ovaries are finely tubular (or flat-banded) and translucent, with inconspicuous branching vessels, and are a light flesh-red color. However, the egg grains are not visible to the naked eye. The spermathecae are flattened and slightly transparent and grayish white or grayish brown in color.

Stage III: Individuals whose gonads are maturing. The gonads are more developed, the ovaries occupy 1/3–1/2 of the entire abdominal cavity, the large blood vessels of the ovaries are obviously thickened, and the grains of eggs stick to each other in clumps. Opaque, slightly white, or pale yellow grains are clearly visible to the naked eye, but when the ovaries are cut open, the grains are difficult to dislodge from the ovarian membrane. The surface of the spermathecae is grayish white or slightly reddish, and no seminal fluid flows out when the spermathecae are squeezed.

Stage IV: Individuals whose gonads will mature. The ovaries occupy approximately 2/3 of the abdominal cavity in volume, and the branching vessels are clearly visible. The egg grains are distinct and rounded. They are easily separated from each other, and sometimes translucent eggs can be seen. The ovaries are orange or orange-red in color. No mature eggs flow from the belly of the fish when lightly pressed. The spermathecae are distinctly enlarged and white in color. There is a small amount of seminal fluid flowing out of the spermathecae when the membrane of the spermathecae is picked or when the belly of the fish is lightly pressed.

Stage V: Individuals with fully mature gonads, about to or in the process of spawning. The gonads are full and fill the body cavity. The ovaries are soft and enlarged, the eggs are large and transparent, and when the ovaries are squeezed or the head of the fish is handheld and slight pressure is applied to the abdomen, the eggs flow out. The spermathecae are developed to their maximum size, milky white, and full of semen. When squeezing the spermathecae or applying slight pressure to the abdomen of the fish, the seminal fluid flows out.

Stage VI: Individuals after egg laying and spermatogenesis. Gonads atrophied, flaccid, and thick; ovaries dark red and significantly reduced in size, occupying only a small portion of the body cavity. The ovarian condyles are thickened. A few mature or small immature egg grains or seminal fluids remain inside the ovary and spermathecae, and the ends sometimes appear bruised.

Depending on the situation and needs of different fish, a stage may also be subdivided into A and B stages, such as the V_A and V_B stages. If gonadal maturity is between two adjacent stages, then the numbers of the two stages can be written, and the maturity stage that the fish is most in can be written in front, e.g., stages III–IV and IV–III. For fish in which gonadal neutrophils mature in batches with multiple outputs, gonadal maturity can be noted based on the development of the gonadal cells that have spawned and those that remain, e.g., stages VI–III, indicating that a portion of the ovary is still in stage III after spawning but has some of the characteristics of stage VI in the appearance of the ovary.

5.1.3.2 Sexual Maturity Factor

In addition to the visual method described above, the maturity coefficient is a measure of gonadal development and is expressed as a percentage of the weight of the gonad compared to the weight of the fish and is calculated as follows:

$$Maturation\ index = \frac{Gonad\ weight}{Net\ weight} \times 100$$

Annual changes in maturity coefficients reflect the degree of gonadal development, and in general, the higher the maturity coefficient, the further along the gonadal development is. The frequency of distribution of the maturity coefficient can be used as one of the indicators to distinguish between sexually mature and sexually immature fish. However, it is inappropriate to use it as the main basis because of the significant differences and fluctuations in maturity status.

It is generally accepted that the closest relationship exists between the early and late initial sexual maturity of fish and their body length. This is due to the ability of fishing to change the structure of fish populations, as well as the influence of factors such as resource density, deteriorating nutritional conditions, and the abundance of aggressive fish. Thus, the age of sexual maturity of fish varies because nutrition before sexual maturity is mainly used for growth in terms of body length. For example, early sexual maturity occurs in the Eastern Sea striped bass due to increasing fishing effort. From a minimum anal length of 238 mm (then sexual maturity IV) in the early 1960s to a minimum anal length of 180 mm at sexual maturity in the late 1970s, a decrease of 22 mm occurred between the two periods.

The general pattern of variation in gonadal maturation coefficients in fish is generally as follows: (1) each fish species has its own maturation coefficient; the coefficients vary from species to species. (2) The individual maturation coefficients are highly variable and increase slightly with age and body length; this indicates that the gonadal maturation coefficients of large and small individuals at the same stage can differ by a factor of 2. (3) The maximum maturation coefficients of batch-spawning fish are generally slightly smaller than those of single-spawning fish. If changes in maturity coefficients are expressed as curves, then batches of spawning fish remain at high levels for longer than single-spawning fish, and the maturity coefficients of single-spawning fish decline sharply after spawning, resulting in a steep downward slope of the postspawning curve. (4) During the transition from sexual immaturity to sexual maturity, the maturity coefficient rises gradually because the weight of the ovaries increases more rapidly than the weight of the fish. When the ovary is in stage II for a long time, the maturation coefficient does not change substantially even if the length and weight of the fish increase. (5) In most Northern Hemisphere fish, the maturation coefficient reaches its maximum in spring, is minimal in summer, and begins to rise again in autumn, and the maximum maturation coefficient for fish spawning in autumn and winter occurs in autumn.

5.1.3.3 Gonadal Index

The gonadal index is also an important element when studying fish reproductive habits. The analysis of gonadal indices allows us to understand the extent of gonadal development in fish and the relationship between gonadal development and fish length, body weight, etc. In general, gonadal indices include the following calculations:

Spermatozoa body index ($J_{T,s}$) = Spermatozoa weight/body weight × 100
Spermatozoa length index ($J_{T,L}$) = Spermatozoa length/body length × 100
Ovary body index ($J_{NL,s}$) = Entangled ovarian gland weight/body weight × 100
Ovary length index ($J_{NG,L}$) = Entangled gland length/body length × 100

The relationship between gonadal index and body weight and mantle length of *Nototodarus* in New Zealand is shown in Fig. 5.2, where the relationship between spermatogonial body index and body weight in males is given by the equation $J_{T,S} = 1.18 \times 0.0011\ W$ ($R = 0.7891$). In general, the sexual maturity of individuals with a spermatogonial body index less than 1.40 is stage I; those between 1.4 and 1.7 are stage II; those above 1.7 are stage III; and some individuals have reached stage IV when the spermatogonial body index is 1.9 (Fig. 5.2).

In this study, the relationship between the spermatogonial length index and carcass length

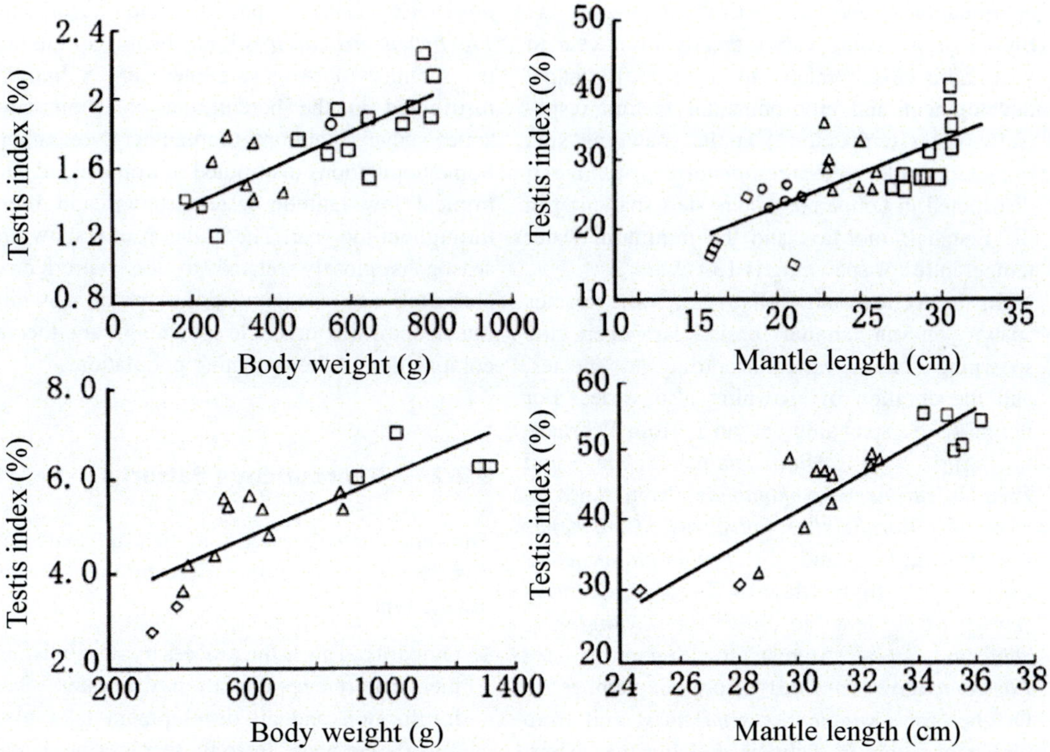

Fig. 5.2 Relationship between various gonadal indices and body weight and length (Chen 1991)

in males is given by $J_{T,L} = 4.740 \times 0.9321$ L (R = 0.7707). Individuals with a general spermatogonial length index value less than 26 are in stage I of gonadal maturity; those with an index value between 26 and 30 are in stage II of sexual maturity; and those with an index value above 30 have essentially reached stage III and very occasionally may reach stage IV (Fig. 5.2).

The relationship between the ovary index and body weight in females is given by the equation $J_{NG,S} = 2.942 \times 0.0030$ W (R = 0.8084). Individuals with an ovary index value less than 3.5 generally are in stage I of sexual maturity; those with an index value between 3.5 and 6.0 are in stage II of sexual maturity, and those with an index value greater than 6.0 have reached stage III (Fig. 5.2).

The relationship between the gonadal length index and carcass length in females is given by $J_{NG,L} = 31.894 \times 2.430$lL (R = 0.8986). Generally, individuals with an ovary length index value less than 30 are in stage I of gonadal maturity; those with an index value between 30 and 50 are in stage II of sexual maturity; those with an index value above 50 are in stage III; and females in stage IV of sexual maturity have not been captured (Fig. 5.2).

5.2 Reproductive Habits

5.2.1 Fish Breeding Period

The breeding time and length of a population are important attributes of a species. Breeding times and spawning grounds are relatively stable and regular. The reproductive period of a fish varies by species and by different populations within a species, each of which engages in spawning activities during a certain season to ensure the continuation of the race. At the same time, the timing of spawning varies greatly from year to year and is closely related to the state of gonadal development and environmental factors (especially water temperature) at the spawning site. For example, *Fenneropenaeus chinensis* distributed in temperate waters start spawning in the first half of May, and the minimum water temperature for spawning is 13 °C.

In the Yellow and Bohai seas, some species spawn on an annual basis. However, the spawning season varies according to species, and the duration of spawning also varies. For flounder, the spawning season is from February to April for *Microstomus achne* and *Pseudopleuronectes yokohamae*; from April to May for *Paralichthys olivaceus*, *Cleisthenes herzensteini*, and *Pseudoplouronectes herzensteini*; from May to June for *Zebrias zebra*; from June to July for *Cynoglossus robustus*; from August to September for *Pleuronichthys cornutus*; from September to October for *Cynoglossus semilaevis*; and from November for *Platichthys bicoloratus*. Thus, flounder species spawn almost every year. However, in general, there are two peak spawning periods for fish in the Yellow Sea and Bohai Sea, and one is in spring and summer; i.e., the warm spawning season results in the most abundant spawning. The other peak is in autumn with the cool spawning period, but fewer types and numbers of species spawn in autumn than in spring and summer. The remaining species spawn in the summer and winter seasons, with the former being mostly warm-water species and the latter being cold-water species. In addition, the spawning period of the same species varies from place to place, as in the case of *Hemiculter leucisculus*, which spawns from May to June in the Yellow and Bohai seas, from February to April along the Fujian coast, from November to January in the northern South China Sea, and from May to June in the Japanese archipelago.

Species distributed at higher latitudes generally spawn once a year. The spawning time varies considerably between ecological types, with *Pseudopleuronectes yokohamae* starting to spawn in early-mid March and *Lateolabrax japonicus* in late September and warm-water *Trachypenaeus curvirostris* spawning in the hotter months of August–September. Salmonids distributed in the boreal and cold temperate zones spawn in autumn (September–November). Some populations distributed in tropical and subtropical low latitude waters spawn and breed throughout the year. The duration of a spawning season is closely related to the reproductive characteristics of a population, type of spawning (batch spawning or single spawning), and the age composition of the spawning population.

5.2.2 Reproductive Patterns

The reproductive patterns of fish are extremely diverse and can be summarized by the three following types:

1. Oviparity. This is the process by which fish lay their mature eggs directly in the water, where fertilization and all development take place outside the body. In some species, the parents do not protect the eggs laid. Since the eggs are not protected by the parents, there is a large possibility that they may be consumed by enemies or exhausted, so these fish have a high reproductive capacity to ensure offspring survival; most marine fish belong to this category. For example, of marine fish, *Mola mola* produce the largest number of eggs, as many as 300 million. Other fish have egg-guarding behaviors that protect their eggs from being eaten by enemies. The way in which egg protection is carried out varies considerably. Some species, such as *Pungitius sinensis*, *Belontiidae*, *gouramies*, and *Channa argus*, dig nests in plants, rocks, or sandy soil to lay their eggs, which are then guarded by males (and occasionally by females) until the young hatch. Some species, such as *Apogon*, brood their eggs in their mouths until the young hatch; some species gestate their eggs on the abdomen. Additionally, some Elasmobranchii (e.g., *Heterodontus*, *Scyliorhinus*, *Apriondon*, and Rajiformes) are oviparous, but the eggs are fertilized in vivo in the female reproductive

tract and then discharged into the water without a second fertilization to complete development.

2. Ovoviviparity. This mode of reproduction is characterized by eggs that not only are fertilized in vivo but also develop in the female reproductive tract; however, the developing embryo is dependent on its own yolk for nutrition, and the mother does not supply the embryo. The embryo is dependent on the mother for respiration. This mode of reproduction is practiced by *Mustelus manazo*, *Lamna nasus*, *Dasyatis*, and bony fish such as *Gambusia affinis*, *Ditrema temmincki*, and *Xiphophorus hellerii*.
3. Viviparity. Some fish have a similar reproductive process as that of mammals, called viviparity. The fetus is in a cyclic relationship with the mother and depends on her for its nutrition in addition to its own yolk, and an example is *Mustelus griseus*.

5.2.3 Types of Spawning

The developmental status of eggs in the ovaries of different species varies considerably, with some exhibiting synchrony and others exhibiting asynchrony, reflecting different species spawning rhythms and thus forming different spawning types. Fish spawning type, which determines the nature of resource replenishment, is therefore closely related to the types of fluctuations in fish populations. The spawning type can be classified according to the composition of the egg diameter and the number of spawning events: (1) single-peaked, single spawners; (2) single-peaked spawners, spawning several times; (3) double-peaked, batch spawners; (4) multipeaked, single spawners; and (5) multipeaked continuous spawners. Spawner type is usually determined based on the frequency distribution of the egg diameter composition within stages III–VI ovary and its variation. Since eggs that have developed to a certain size in the ovary (e.g., oocytes with yolk in stage IV) may still be absorbed, it is sometimes difficult to confirm the ovary stage using the egg diameter frequency method alone, and therefore, histological sectioning is required to confirm it.

For example, in offshore China, based on observations of ovarian histology during the reproductive season, Yellow Sea *Clupea pallasii* and *F. chinensis* are considered to be single-peaked, single spawners; Bohai Sea *Portunus trituberculatus* are bimodal spawners, spawning twice; *Trichiurus lepturus* in the East China Sea and *Upeneus bensasi* in the South China Sea are bimodal, batch spawners; *Saurida tumbil* in the northern South China Sea are multipeaked spawners, spawning several times; and *P. major* are multipeaked, continuous spawners.

5.2.4 Types of Spawning Communities

Length, age composition, and sex ratio vary for different groups of fishery resources, and even for the same group, there are differences during different stages of exploitation. A spawning population usually consists of two major components: (1) the population that has spawned in the past, called the residual spawning population, and (2) the population that is sexually mature for the first time, called the supplementary spawning population. Therefore, studying the composition of a spawning population requires determining the ratio of the remaining spawning population to the supplementary spawning population, in addition to studying the body length, age, and sex ratio.

The reproductive groups of fish are usually classified into three types.

The first type is I $RS = 0$ and $SS = SP$.

The second type is $RS > 0$, $SS > RS$, $SS + RS = SP$.

The third type is $RS > 0$, $SS < RS$, $SS + RD = SP$.

where RS denotes the remaining spawning population, SS denotes the supplementary spawning population, and SP denotes the spawning population.

Fishery resource groups that fall into the first category are short-lived fish and crustaceans, such as *F. chinensis*, *Acetes chinensis*, *P. altivelis*, *Hemisalanx prognathus*, *O. keta*, etc., and most cephalopods, which generally die after their first spawning event.

The fishery resource groups that fall into the second category are medium-lived fish, such as *Trichiurus lepturus*.

The fishery resource groups that fall into the third category are long-lived fish and cetaceans, such as *L. crocea* and *Balaenoptera physalus*.

However, the reproductive group to which a fishery resource belongs is relative. Overfishing has a significant impact on the second and third types of fishery resources because fishing activities tend to target remaining spawning populations and reduce the remaining population as fishing intensity continues to increase. When overfishing occurs, spawner type tends to change as well, as in the case of *L. crocea* population, which is overfished and has a decreasing residual portion of the population and a gradually increasing complementary portion; this scenario has caused a gradual conversion of the reproductive population from type III to type II.

5.3 Fecundity Definition and Fertility Measurement Methods

5.3.1 Definition of Individual Fecundity

Fecundity is the absolute or relative number of eggs that a female may expel in a reproductive season. However, since it is often difficult to measure this in research studies, the total number of eggs carried in ovaries equivalent to those in a stage III or higher ovary or their relative number is often used instead.

The fecundity of fish can be divided into individual absolute fecundity and relative fecundity. Absolute individual fecundity refers to the number of eggs that a female may expel in a reproductive season. In practice, two related terms are often encountered: fecundity, which refers to the number of eggs visible in the ovary on the eve of spawning during the process of maturation, and spawning capacity, which refers to the number of eggs that will be produced or has been produced. The actual quantitative values of the two differ; e.g., the spawning stock of small croaker is approximately 90% of the pregnant spawning stock. In terms of definition, "spawning capacity" is closer to "absolute fecundity."

However, in practice, egg counts are mostly determined by the weight sampling method, and the calculation criteria are generally determined by the egg diameters during maturation in stages IV–V ovaries, such as 0.16–0.99 mm *L. crocea* eggs, 1.10 mm or greater for *C. pallasii*, and 0.35 mm or greater for *Thamnaconus modestus*. Calculating the absolute fecundity in this way is similar to calculating "egg carrying capacity." In addition, "absolute fertility" is actually a relative value. The extent to which this relative value is close to the actual individual absolute fecundity depends on the extent to which spawning type is studied, i.e., the criteria for classifying the eggs to be produced, the spawning batch, and the percentage of eggs that may be absorbed.

Individual relative fecundity is the ratio of absolute fecundity to body weight or length, i.e., the number of eggs that may be discharged per unit weight (g) or unit length (mm) of a female in a reproductive season. Relative fecundity is not constant and to some extent varies according to changes in living conditions or growth status. Therefore, it is an important indicator of the reproductive capacity of individuals in a population and can be used for not only comparing different populations of a species but also comparing interspecific differences in the level of value added per unit weight or body length. As described in Table 5.2, there are significant differences in the fecundity per unit weight of some important fishery species in the Yellow Sea and East China Sea.

The specific calculation formula is as follows:

$$\begin{aligned} &\textit{Absolute fecundity} = \\ &\textit{the number of oocyte in n gram samples} \\ &\times \frac{\textit{Ovary weight}}{\textit{n gram}} \end{aligned}$$

$$\textit{Relative fecundity} = \frac{\textit{Absolute fecundity}}{\textit{Body lengt or Net weight}}$$

Table 5.2 Comparison of the relative fecundity of individuals of some species in the Yellow Sea and East China Sea

Type	Number of eggs per unit weight (grains/g)
Liaodong Bay *L. polyactis*	171–841
East China Sea *L. crocea*	268–1006
Yellow Sea *C. pallasii*	2 10–379
East China Sea *Trichiurus lepturus*	108–467
East China Sea *Thamnaconus modestus*	674–2490

Chen (2014)

5.3.2 Patterns of Variation in Individual Fish Fecundity

The fecundity of fish has been shown to vary with body weight, body length, and age. For example, in the case of Chinese offshore *Trichiurus lepturus*, the absolute fecundity of individual *Trichiurus lepturus* increased with length and body weight, and the increase in fecundity increased gradually with length and body weight; i.e., absolute fecundity increased as a power function of anal length and body weight (Fig. 5.3). The absolute fecundity of *Trichiurus lepturus* has also been found to increase more significantly with body weight than with length; for example, the fecundity was essentially the same between a length of 210 mm and a body weight of 150 g, but then, the increase in fecundity according to body weight was gradually greater than the increase according to body length. The study concluded that the absolute fecundity of individual *Trichiurus lepturus* is most closely related to body weight, followed by fish length and then age.

The pattern of change in the relative fecundity r/L of individual *Trichiurus lepturus* was the same as that of absolute individual fecundity, which increased according to anal length, body weight, and age. The relationship between r/W and anal length and body weight showed an irregular undulating curve, indicating that r/W did not change significantly with increasing anal length or body weight. The first absolute fecundity value of *Trichiurus lepturus* also increased according to length, body weight, or age, but the first absolute fecundity was less than the second absolute fecundity (Fig. 5.4a, b). For example, the first absolute fecundity of *Trichiurus lepturus* in the western Taiwan Strait ranged from 15.3 to 117.6 kilos, with a mean of 37.4 kilos; the second absolute fecundity ranged from 18.4 to 156.6 kilos, with a mean of 57.1 kilos.

5.3.3 Mechanisms Regulating Individual Fecundity in Fish

The pattern of change in fish fecundity is the most important pattern of fish population variability. In parallel with changes in food security, changes in species and population fecundity are regulated automatically through changes in material metabolism affecting the degree of population proliferation and controlling population size to adapt to food security.

1. Variation in fecundity relates fish age and size. Fecundity in most fish is more closely correlated with fish weight than with body length, which in turn is more closely correlated with body length than with age.

After a fish reaches the age of sexual maturity, fecundity increases as the fish grows, until it begins to decrease at an advanced age. The relative fecundity of younger age groups is generally the greatest, and older individuals do not reproduce every year. This scenario occurs because the first reproducing individuals have the smallest eggs and higher relative fecundity, which generally increases more slowly over a longer period of time thereafter as the fish grow. The increase in reproduction for older fish is due to an increase in individual senescence, relative fecundity (including absolute fecundity), the number of absorbed egg grains, and changes in the environment where the population is located, resulting in the occurrence of fish that do not spawn in a given reproductive season.

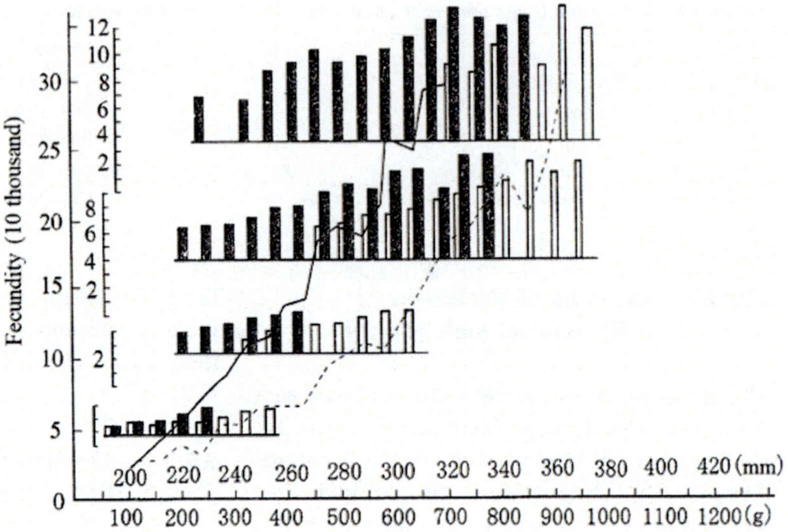

Fig. 5.3 Relationship between individual fecundity and fish length, body weight, and age of *Trichiurus lepturus* in East China Sea (Chen 2014) The black boxes in the figure indicate body length, and the white boxes indicate body weight

2. Variation in fecundity depends on the rate of prey supply. Fish fecundity can be clearly divided into two periods; the first period involves reproductive epithelial growth, when the total individual fecundity possessed by a population is formed. The second period involves effective fecundity formation, where fecundity and yolk accumulation in eggs are

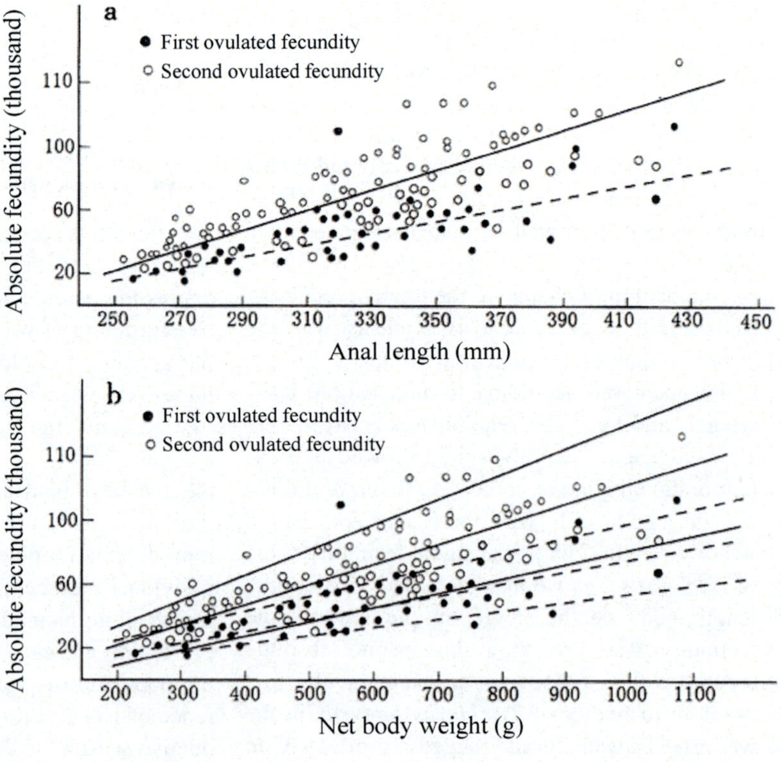

Fig. 5.4 Relationship between individual absolute fecundity and anal length and net body weight of *Trichiurus lepturus* in the western Taiwan Strait (Du et al. 1983) (**a**). Relation to anal length. (**b**). Relation to net body weight

closely related to changes in external food security. Thus, interannual variation in fish fecundity is related to prey conditions during the prereproductive prey season.

The main way to regulate fertility within the same group is to accelerate growth when prey is plentiful; the fatter the fish are, the better the oocytes develop, the more eggs there are, and the more fertility increases. In contrast, in years when there are low levels of prey, some of egg cells shrink and are absorbed, and fertility is reduced.

3. Individual fecundity varies with fish growth. Individual fecundity varies with the shift in the life span of a fish and can generally be divided into three phases, namely, the period of fecundity growth, the period of active fecundity, and the period of fecundity decline. During the reproductive growth period, fecundity increases rapidly; during the active period, the growth rhythm of fecundity is generally more stable, but fecundity reaches its maximum; during the fecundity decline period, the growth rate of fecundity decreases. For example, the fertility of *L. crocea* in Daiquoyang, Zhejiang Province, is low at the ages of 2–4 and moderate at the age of 5, which is the period of fecundity growth when reproductive activity begins; fish at the ages of 5–14 years old are in the active fecundity period, which increases with age; after approximately 15 years of age, fecundity gradually decreases, and the period of fertility decline occurs, which is a reflection of gonadal function decline in the organism.
4. Differences in the fecundity of different populations of the same species. The fecundity of populations of the same species living in different environments is different, and the greater the differences in the environments in which different populations live, the greater the differences in their fecundity are. For example, two populations of *C. pallasii* from North Pacific waters and North Atlantic waters have completely different fecundity levels.

The fecundity of fish of the same species living in the same sea that are of the same size or age may vary if they reproduce at different times. For example, the fecundity of the spring reproductive population of the greater amberjack off Zhejiang is higher than that of the autumn reproductive population.

For similar species of marine fishes, in comparison with those in other regions, species with a southerly distribution are characterized by higher fecundity. The increase in fecundity of these species is achieved by increasing the number of eggs per batch. Thus, similar species exhibit an increase in fecundity from high to low latitudes, which is evident in studies of species that do not batch spawn.

5.3.4 Methods for Measuring Individual Fish Fecundity

The fecundity of various fish species is highly variable. For example, cartilaginous fish such as *Heterodontus japonicus* and *Galeus* spp. only lay 2–3 eggs, while the triggerfish can lay 300 million eggs. Fish that do not protect their eggs after spawning and are more affected by enemies and the environment generally have a larger egg carrying capacity, such as *P. major*, which on average spawns approximately 1 million eggs, with the highest amount of eggs being 2.34 million for *P. major*, 2.9–7.2 million for *Mugil cephalus*, and 7–15 million for *Anguilla japonica* off the coast of Fujian. Usually, marine fishes are more fertile than freshwater fishes and anadromous fishes; those that produce floating eggs are the most fertile, followed by those that produce sunken eggs, and those that reproduce and then carry out conservation or oviparous births are the least fertile.

There are also various methods of counting eggs, including counting, weight proportional, volumetric, and Reibish methods. The egg count method is mostly used for small numbers of large eggs, such as those of salmon, trout, and catfish species, and the weight proportional method is usually used for other fish.

1. Weight proportional method. After a biological assay has been performed, the ovary is removed and weighed; then, a sample of 1 g or less is taken from the whole ovary according to the size of the grains; and the number of grains is calculated. If the size of each part of the ovary varies, then some samples should be taken from different parts of the ovary, and their average value should be calculated (e.g., 0.2–0.5 g each of the anterior, middle, and posterior parts); then, the number of egg grains contained in the whole ovary should be deduced by the proportional method. The formula is the following:

$$FE = \frac{Ww}{Ws} \times fe$$

where FE is the absolute fecundity (grains), fe is the number of eggs in the sample (grains), W_w is the weight of the whole ovary (g), and W_s is the weight of the sample (g).

The relative fecundity of a fish is the number of eggs carried per unit of body length or body weight. When calculating individual fecundity, care should be taken to use stage IV ovaries rather than stage V ovaries, which may have had some eggs extruded from the body. It is important that the part of the ovary used for the calculation be chosen for its representativeness. In addition, fresh specimens are preferable for the calculation of fecundity, or if this is difficult, then specimens immersed in a 5% formaldehyde solution may be used.

2. Volume ratio method. The ratio of the volume part of an ovary to the volume of the whole ovary is multiplied by the volume of eggs contained in the part of the ovary to determine the total number of eggs carried. The drainage method is used to determine the volume of the ovary and eggs:

   The formula is the following:

$$E_i = \frac{V}{U} \times e$$

where E_i is the total ovary egg carrying capacity, V is the volume of the ovary, U is the weight of the ovary part (g), and e is the number of eggs contained in the part of the ovary.

Selected partial ovaries were macerated using Simpson's solution (Simpson), and all eggs were separated, aspirated, and counted. However, e often varies depending on the part of the ovary taken. Therefore, several partial egg masses should be taken from different parts of the ovary, and e should be averaged.

5.3.5 Fertility of Fish Stocks and Methods for Its Estimation

Individual fecundity sometimes does not accurately reflect the actual reproductive capacity of a population due to variations in growth, age at sexual maturity, population composition, number of parents, etc., and the fecundity of a population needs to be studied. Population fecundity refers to the total number of eggs that may be produced by all females in a reproductive season. A sound estimation method is still lacking, and an approximate formula for estimating population fecundity is usually the following:

$$E_p = \sum N_x \times F_x$$

where E_p is population fecundity, N_x is the number of females likely to spawn in an age group (number of tails), and F_x is the average individual fecundity (number of eggs) for the same age group.

Where fecundity per unit weight is relatively stable, population fecundity can also be expressed as the product of individual relative fecundity and the biomass of spawning females. For example, in the case of Yellow Sea *C. pallasii*, fish that are 2 years old have essentially reached full sexual maturity, the sex ratio of the spawning population is close to 1:1, and their individual fecundity varies with age and is tabulated and calculated as follows (Table 5.3) in conjunction with the year-by-year generation analysis:

As seen in Table 5.3, the population fecundity of Yellow Sea *C. pallasii* fluctuates greatly depending on the year, ranging from 4 to 2.27 billion grains, which is influenced by the strength of

Table 5.3 Fecundity of the Yellow Sea *C. pallasii* population

Age	Mean fecundity (ten thousand)	Year 1969		Year 1970		Year 1972	
		Spawning female* (ten thousand)	Stock fecundity (hundred million)	Spawning female* (ten thousand)	Stock fecundity (hundred million)	Spawning female* (ten thousand)	Stock fecundity (hundred million)
2	3.07	337.50	10 246.13	6 487.15	19 915.55	70 598	216 735.86
3	4.90	5 724.55	28 050.30	1 996.60	19 783.34	493.8	2 419.62
4	5.45	302.45	1 648.35	3 215.05	17 522.02	585.25	3 189.61
>4	5.43			246.90	1 340.67	831.8	4 516.67
Total		9 364.50	39 944.78	11 945.70	48 561.58	72 508.85	226 861.76

Note: Spawning female for each year = (number of females likely to spawn in an age group N_x − catch of females likely to spawn in an age group C_x)/2

Chen and Liu (2017)

the dominant generation of the spawning population; e.g., the fish that were 2 years old in 1972 or the 1970 generation were very fertile, resulting in a sharp increase in population fecundity in that year. This fluctuation is also found in other species, and the range of variation varies depending on the species.

The magnitude of population fecundity varies considerably among species, and this variation reflects the adaptive nature of species and populations to environmental changes at spawning grounds. In general, marine fish are more fertile than freshwater and anadromous fish; migratory and anadromous fish are more fertile than sedentary species. Fecundity was greatest in floating spawning fish, followed by sinking spawning fish, and least in oviparous fish. Differences in fecundity between populations of the same species are greater as the differences in environmental conditions become greater. Systematic information on the interannual variation in the fecundity of populations is important for exploring the dynamics of their replenishment.

5.4 Reproductive Strategies for Fish

5.4.1 Reproductive Strategies

Reproductive strategies, the different reproductive approaches adopted by individuals of a species to achieve a higher fitness for reproduction, are evolutionarily stable strategies for variability in reproductive behavior.

Reproductive strategies are relatively stable rules, and the fitness of different rules is the result of a combination of natural and sexual selection during evolution, reflecting the different resource use patterns and tolerance to environmental fluctuations of individual organisms, as well as their adaptability to their living environment and maximization of reproduction. Within a relatively narrow range of habitats, individual organisms must maximize reproductive value and optimize reproductive effort choosing between resource acquisition and reproductive continuation (Fig. 5.5). The environment here includes the biotic environment, such as prey and number of available mates, and the abiotic environment, such as habitat area, temperature, and light.

5.4.2 Types of Reproductive Strategies for Fish

The reproductive strategy of a fish is life history specific. Different fish appropriately regulate and arrange each resource allocation and their reproductive behavior according to their life cycle, while appropriate resource allocation is determined by the reproductive response adopted and its accompanying environmental state, such as the biological environment (competitors, predators, prey, parasites, etc.) and the physical environment (temperature, humidity, effective oxygen, salinity, etc.). Thus, each individual fish can be simply

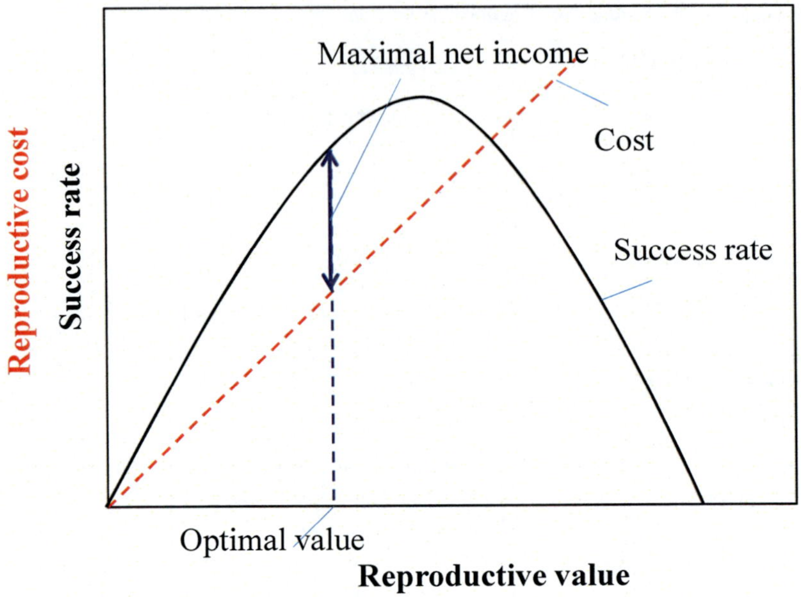

Fig. 5.5 Schematic diagram of reproductive value and reproductive effectiveness (Chen and Liu 2017)

understood as a simple input-output system (Fig. 5.6): the acquisition of material resources through appropriate feeding activities and efficient conversion into body energy, the appropriate consideration and choice of resource allocation between individual growth and reproduction for each reproductive behavior, the amount of each resource allocated to each egg (or pup), etc. The most appropriate reproductive response is then chosen to successfully reproduce.

Thus, the reproductive strategy of fish can be defined in two ways: a strategy defined by the classification of spawning patterns, which can be divided into single reproduction (semelparity) and multiple reproduction (iteroparity), and a strategy defined in terms of the design of ecological responses, which can be divided into the *r*-strategy and *K*-strategy.

In terms of spawning patterns, single reproduction involves the maturation of individual gonads, the energy of the organism is used to reproduce and spawn at once, and then, the fish dies after reproduction, with no postspawning surplus population. During the breeding season, individual fish produce all of their eggs and then die (Fig. 5.7a). Multiple reproductions involve the maturation of individual gonads and the use of only part of the body's energy for reproductive spawning or the use of reproductive inputs derived entirely from prey ingestion, and the fish survives after reproduction and continues to reproduce and spawn the following year or in subsequent suitable seasons. During the breeding season, the spawning behavior of an individual fish may be either a single-spawning event or a batch-spawning event. The former is when all the eggs formed when the ovary matures are expelled, and then, the ovary becomes dormant to await the next spawning season (Fig. 5.7b). In the latter, eggs are expelled in batches during breeding, which may vary from one to several days apart, and this type of oviposition is closely related to fecundity: if fecundity is deterministic, then an individual fish expels the eggs due in batches and the oviposition ends without new egg replenishment (Fig. 5.7c); if fecundity is uncertain, then an individual fish continuously produces new eggs in the ovary and oviposits in successive batches of oviposition (Fig. 5.7d).

In terms of ecological response design, individuals within *r*-strategy populations often devote more energy to reproduction and less to growth and metabolism, enhancing their

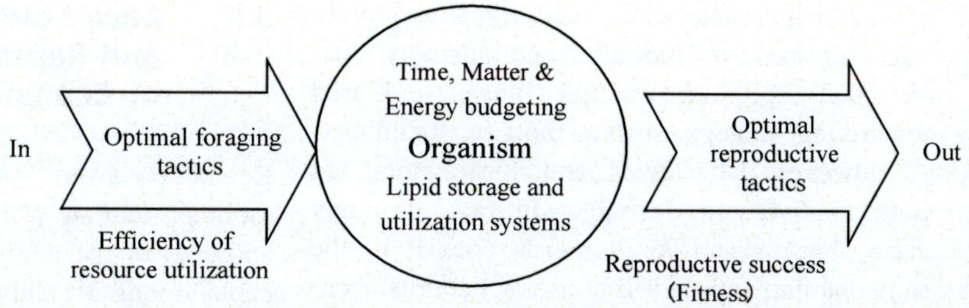

Fig. 5.6 Schematic diagram of the input-output system of an organism (Pianka 1976)

competitive ability, i.e., high endogenous natural growth rates with *r*-strategy populations. Their life histories reflect greater fitness in uncertain or fluctuating habitats and exhibit significant environmental and population fluctuations and high resource replenishment recovery. The attributes of such fish are early sexual maturity, rapid growth rates, small size, high mortality, and short life spans. Many single-spawning fish species, such as salmon and oceanic ommastrephids, favor this strategy.

In contrast, in comparison to those in *r*-strategy populations, individuals in *K*-strategy populations often use more energy for a variety of activities other than reproduction, such as growth and metabolism, enhancing their competitive ability, i.e., environmental accommodation in *K*-strategy populations is superior, and their life histories have a high fitness in stable habitats, exhibiting lower population fluctuations and greater environmental resistance and slow recovery from resource replenishment. The attributes

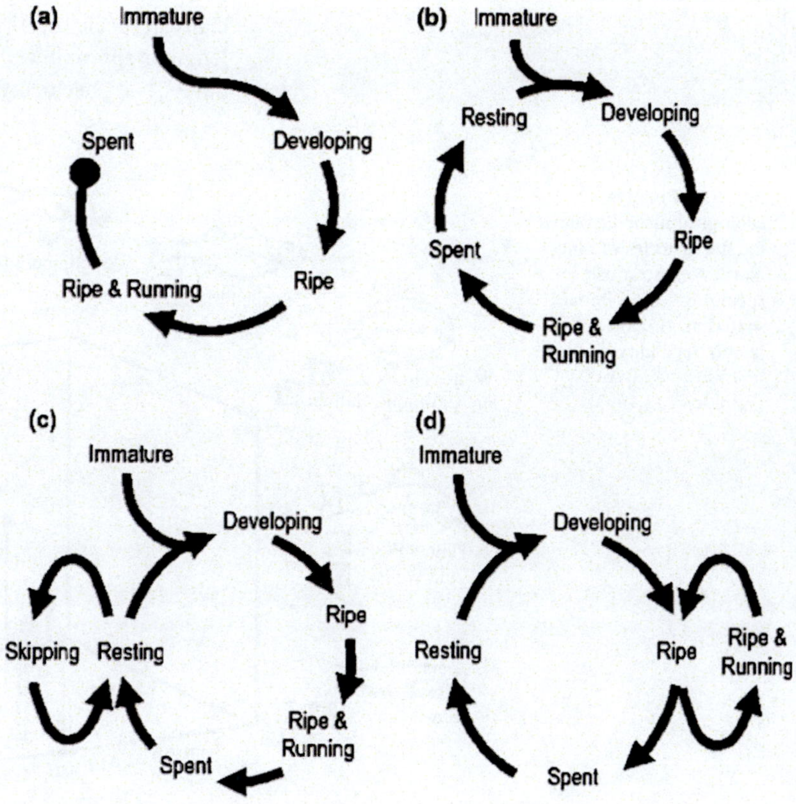

Fig. 5.7 Schematic diagram of fish reproductive strategies based on spawning patterns (McBride et al. 2015)

of such fish are late sexual maturity, slow growth rate, large size, low mortality, and longevity. Species that reproduce multiple times are biased toward this strategy, such as most scleractinians.

However, *r*-strategies and K-strategies are both relative and desirable. In fact, in many cases, both *r* and *K* strategists coexist in the same habitat, and their life history patterns occur on the *r*-*K* spectrum band, with relatively few *r* and *K* strategists being particularly typical and many species exhibiting relative ecological positions at intermediate levels (Fig. 5.8). In addition, for the same species that also exhibit a variety of life history characteristics depending on their habitat, such as the North American *Lepomis gibbosus*, groups living in Little Round Lake and Warren Lake exhibit more of an *r*-strategy, while groups living in Beloporine Lake are biased toward a *K*-strategy.

5.5 Case Study: Sexual Maturation and Reproductive Habits of Cephalopods

Cephalopods are ancient and higher marine mollusks that are widely distributed in all oceans and seas of the world. The life cycles of extant cephalopods are short, with most medium and small cephalopods living 1–2 years and some large cephalopods living up to 8–10 years, while nautiluses generally live 5–10 years or even more than 20 years. With the exception of *Nautilus*, all genera have a single developmental reproductive maturity stage, and no additional germ cells will appear in the sperm/ovary after gonadal maturity, a reproductive adaptation of the species to reproduce once in its lifetime, with sensitive environmental adaptations.

Cephalopod species are sexual dimorphism, with loliginids, cuttlefishes, and a few octopuses (e.g., *Octopus vulgaris*) exhibiting larger males than females, while ommastrephids and other octopuses exhibit larger females than males, especially in species of *Argonauta*, where male mantle length is approximately 1/5 that of females. Males have 1–2 hectocotyli, and the reproductive

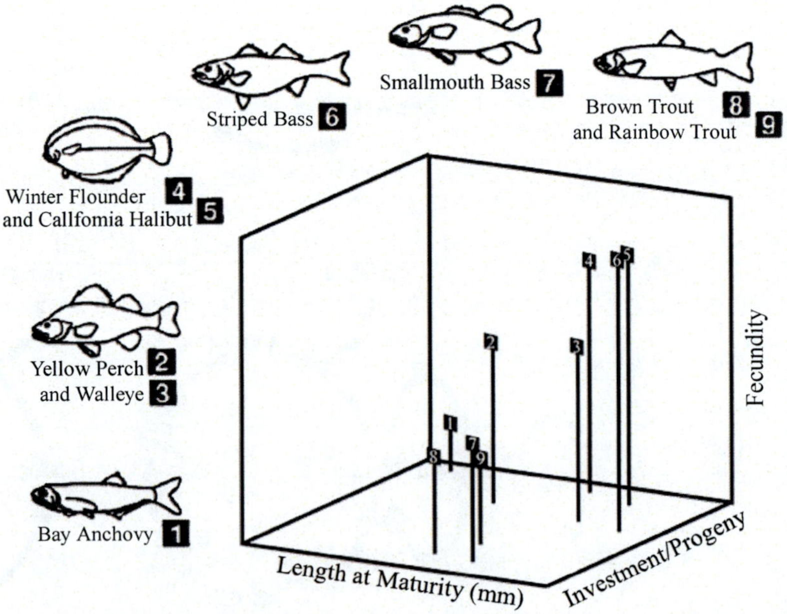

Fig. 5.8 Relative ecological niche layout of the life histories of nine North American fish species based on initial sexual maturation body length, fecundity, and reproductive inputs (Winkle et al. 1993)

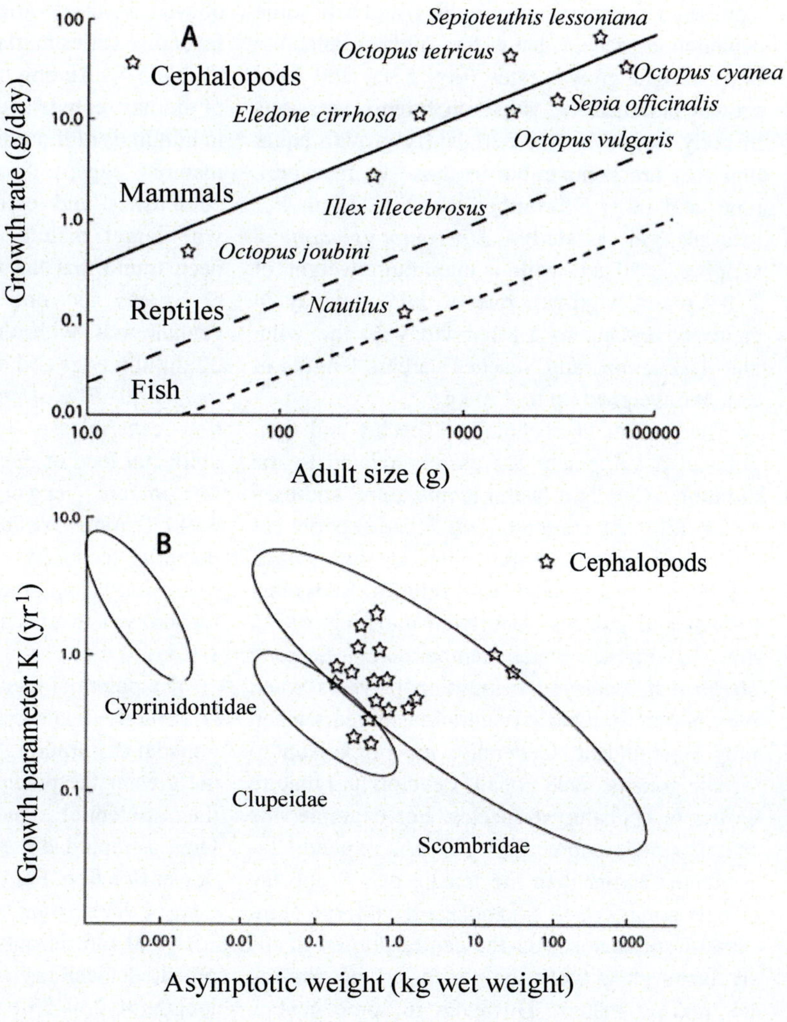

Fig. 5.9 Growth of cephalopods (Boyle and Rodhouse 2005) (**a**). Relationship between mean growth rate and adult size for cephalopods, mammals, reptiles, fish, etc. (**b**). Relationships between regression growth rate K and body weight that fit von Bertalanffy growth equations for cephalopods, fish, etc.

system consists of testicle, spermatophoric glands, spermatophoric duct, Needham's sac, and terminal organ; the female reproductive system consists of ovaries, nidamental gland (and accessory nidamental gland), oviduct, oviductal gland, etc. Females of some genera also have special photophore at arm tip; only some female *Enteroctopus megalocyathus* have been found to be hermaphroditic due to environmental conditions. The ratio of females to males is generally close to 1:1; for example, the number of females and males in benthic octopuses and cuttlefishes is comparable, while the ratio of females to males in pelagic *Loligo* species is slightly more than that of females; however, the overall ratio is close to 1:1, and the sex ratio does not change significantly throughout the reproduction process.

5.5.1 Growth and Sexual Maturity

Cephalopods grow rapidly and are the largest genus of invertebrates in terms of body size.

Compared with mammals, reptiles, and fish, some cephalopod genera have the highest absolute body weight growth rates (Fig. 5.9a) and have growth parameter *K* values in fitted regressions of body weight in von Bertalanffy growth equation that are comparable to those of mackerel, tuna, and other Schromberidae (Fig. 5.9b). For example, in a study, *Enteroctopus dofleini* weighed >50 kg, with a maximum weight of 272 kg and a growth rate of 1.1% bw/day in captivity and up to 1.8% bw/day in the wild, and *Architeuthis* spp. reached carcass lengths of 2 m and weighed up to 450 kg.

The growth of cephalopod species and their gonadal development are closely related to the duration of sunlight, water temperature, and prey availability; e.g., *Sepiella japonica* experiences early maturation and spawning in winter when sunlight is short, and it promotes individual growth and induces maturation and oviposition due to higher water temperature; *Eledone cirrhosa* has delayed maturation in years when prey is poor and early maturation in years when prey is abundant. Generally, in comparison to female gonads, male gonads develop and mature earlier in cephalopod species. For example, the male spermatophore of *O. vulgaris* expands 2–-3 months earlier than the female ovary, and the female gonads of *O. tankahkeei* have large numbers of spermatozoa in the central lumen of the oviductal gland when the gonads are developing but not yet mature. However, in some genera, male gonads develop earlier than those of the females, but the maturation of both gonads is simultaneous, as in *I. argentinus*. Male prematurity is an adaptation of the species to the presence of mating between males and females and overall exocytosis of sperm pods during the reproductive season, whereas simultaneous maturation of male and female gonads is a reproductive adaptation to the absence of nasal vesicles or similar structures in the reproductive system.

5.5.2 Fertility Characteristics

In general, the fecundity of cephalopods is estimated mainly on the basis of the counts of oocytes in the ovaries and oviducts. Maximum fecundity varies markedly among different genera (Table 5.4). Among them, the potential fecundity of ommastrephids and loliginids are both counted in hundreds of thousands and millions, with small mature eggs; the potential fecundity of cuttlefishes and octopuses is relatively small, with larger mature eggs. For example, it has been found that the fertility of the jumbo squid *D. gigas* is one of the highest among cephalopods, with a potential fertility as high as 32 million eggs and an actual fertility of approximately 50–70% of potential fertility. Each ovulation can release approximately 80% of the oviductal load of eggs, and the average diameter of mature eggs is only approximately 0.9–1.1 mm. While *Sepia orbignyana* has a potential fecundity of less than 2000, its actual oviposition is approximately 50% of the potential fecundity, and its mature egg size is greater than 1.2 × 3.0 mm.

The potential fecundity of cephalopod species is strongly related to their individual size and gonadal maturation. The larger the individual is, the greater the potential fecundity. For example, the potential fecundity of species of Ommastrephidae is related to carcass length as a power function. For example, the potential fecundity of *Illex coindetii* with a carcass length of 150–160 mm is only 90,000 oocytes, while the potential fecundity of *I. coindetii* with a carcass length of 230–250 mm is as high as 800,000 oocytes. However, as the gonads mature, the potential fecundity of an individual decreases (Fig. 5.10). On the one hand, the potential fecundity of an individual decreases significantly due to the degradation and absorption of oocytes as a direct result of nutrient supply, mutual extrusion during oogenesis, and degeneration of egg-laying ovaries; on the other hand, the potential fecundity of an individual decreases gradually because a mature individual lays eggs in batches and new growth of newborn oocytes ceases. For example, the potential fecundity of immature female gonads of *Sthenoteuthis pteropus* is as high as 17.91 million oocytes; after the gonads are mature, the potential fecundity decreases to 5.8–-15.81 million oocytes.

Table 5.4 Cephalopod fecundity

Genus (taxonomy)	Potential fertility	Actual oviposition	Oocyte size (mm)
Sepioids			
Sepia officinalis	150–4 000	500	1.2 × 3.0
Sepia orbignyana	201–1532	400	6.3–8.3
Sepiola robusta	35–54	–	3.5 × 4.5
Euprymna scolopes	300	–	4.0 × 4.0
Sepietta oweniana	28–236	<160	4.5 × 5.0
Rondeletiola minor	5–460	100	15–30
Loliginids			
Loligo opalescens	10,000–12,000	<4250	1.6 × 2.5
Loligo pealei	3500–6000	21,000–53,000	1.0 × 1.6
Loligo vulgaris	6000–10,000	<7000	2.2 × 2.7
Ommastrephids			
Illex argentinus	75,000–1,200,000	<840,000	0.96–1.04
Illex illecebrosus	200,000–630,000	<100,000	0.8 × 1.0
Todarodes pacificus	320,000–470,000	<500,000	0.7 × 0.8
Dosidicus gigas	<32,000,000	<6,000,000	0.9–1.1
Octopods			
Octopus briareus	300–500	200–500	5.0 × 14.0
Octopus cyanea	<700,000	<600,000	<3.00
Octopus dofleini	18,000–70,000	–	<8.0
Octopus joubini	<321	–	4.0 × 8.0
Octopus maya	3000–5000	1500–20,000	3.9 × 11.0
Octopus tetricus	<700,000	<15,000	0.9 × 2.4
Octopus vulgaris	13,000–634,000	100,000–500,000	1.0 × 2.0
Eledone cirrhosa	2000–54,000	800–1500	2.5 × 7.5
Eledone moschata	187–944	100–500	5.0 × 16.0
Bathypolypus arcticus	20–80	20–80	6.0 × 14.0

Lin (2015)

5.5.3 Reproductive Strategies

Cephalopods are characterized by sensitive adaptations to the marine environment; genera of octopus, cuttlefish, and nautilus are more inclined toward *K*-strategies, while genera of gunny loliginid and ommastrephid squids tend toward *r*-strategies. These genera have various forms of adaptive reproductive habits, such as migratory spawning, seasonal spawning, and specific spawning strategies.

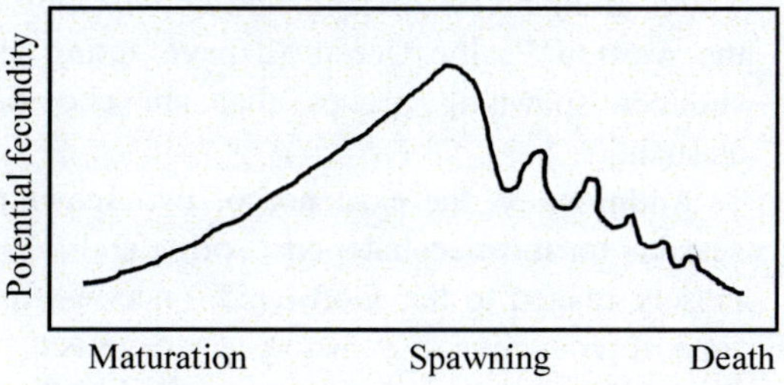

Fig. 5.10 Schematic diagram of the variation in potential fecundity with gonadal maturity for spawning in the *Illex* (Laptikhovsky and Nigmatullin 1993)

5.5.3.1 Migratory Spawning

Cephalopods, with the exception of octopuses, are induced by the temperature of their habitat and gonadal development to migrate from feeding grounds to spawning grounds in the later stages of their lives. For example, during the spawning season, cuttlefishes migrate from deeper shelf waters to shallower coastal waters to spawn, with a suitable water temperature for spawning of 13–15 °C. *Doryteuthis gahi* migrates to spawn twice a year, in spring and autumn, from the

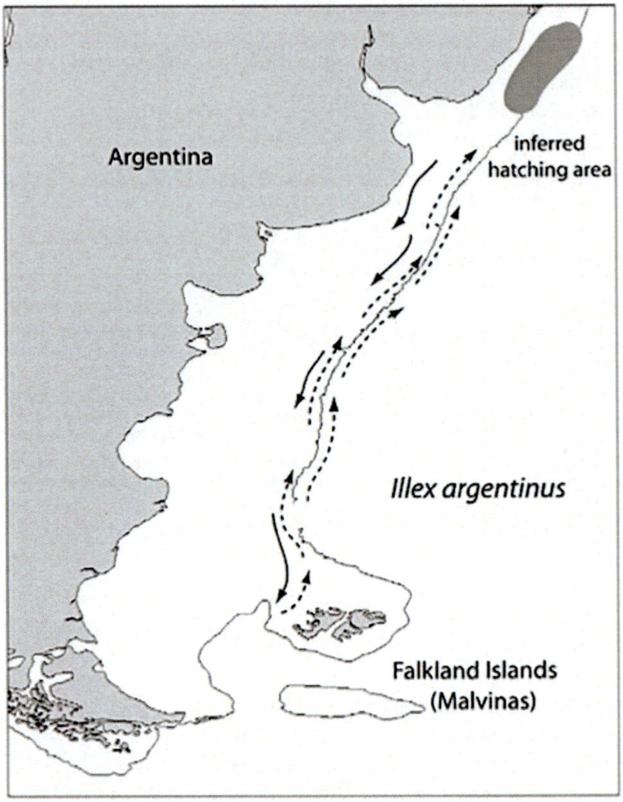

Fig. 5.11 Schematic diagram of spawning migration (dashed arrows) and feeding migration (solid arrows) of the winter spawning population of *I. argentinus* (Boyle and Rodhouse 2005)

deeper shelf edge and shelf slope waters to shallower coastal waters near the Falkland Islands.

In addition, oceanic ommastrephids are typical migratory spawners, and migrations are generally closely associated with currents. For example, the winter spawning group of *Todarodes pacificus* are located in the central and northern East China Sea; the feeding grounds are as far northwest as Hokkaido, Japan; and the spawning migrations are closely associated with the Kuroshio Current and the first branch of the Tsushima Current; and the winter spawning group of *I. argentinus* migrate northward with the Falkland Current from the waters off the Falkland Islands and the Patagonian shelf, spawning near the confluence of the Falkland and Brazilian currents, while juveniles migrate southward to feed and grow (Fig. 5.11).

5.5.3.2 Spawning Season

Similar to fish, all cephalopod genera have a certain spawning season. In general, spring and summer appear to be spawning seasons for most cephalopods, when water temperatures are favorable and prey are abundant. For example, *Sepia esculenta* overwinters in the deeper waters every spring and swims in clusters to breed in shallow waters; *Sthenoteuthis oualaniensis* in the northwestern Indian Ocean, *Nototodarus gouldi* in the waters around New Zealand, and *Loligo edulis* in the western Pacific Ocean all have spring and summer spawning groups that are classified as such.

Additionally, the existence of two spawning seasons for most cephalopods, spring and fall, is closely related to the short-lived, once-in-a-lifetime reproductive life history characteristic of these genera to perpetuate the reproduction of resource biomass. For example, *D. gahi* in the southwestern Atlantic spawn in October–

November for the fall spawning population and in April–May for the spring spawning population, and the difference in water temperature between the two spawning grounds results in a difference of only 2–3 months in the hatching time of fertilized eggs between the two spawning populations. *Ommastrephes bartramii* in the North Pacific are predominantly winter-spring spawners from January to May, with a few spawning individuals in the fall.

In addition, the reproductive season of the same genus of cephalopods can vary depending on their habitat distribution, morphological size, etc., and spawning can even occur throughout the year. For example, cuttlefishes spawn in shallow waters and throughout the year, with a peak spawning period in April–July in western Mediterranean waters; off Senegal and along the Saharan coast, the peak spawning period for large individuals is in January–April, while the peak spawning period for small and medium individuals is in late summer and early autumn. *I. argentinus*, the Buenos Aires-North Patagonia group distributed at 37°–43°S, spawn in winter in continental slope waters; the South Patagonia group, distributed at 43°–50°S, spawn in autumn; the summer spawning group, distributed in continental shelf waters at 42°–48°S, spawn in summer; and the group distributed in Argentine coastal waters, such as those in San Matías Bay, spawn in summer due to the suitable water temperature and prey. *T. pacificus*, distributed along the Pacific coast of the Japanese archipelago, spawn from January to March, those distributed in the central sea of Japan spawn from September to November, those distributed in Japanese coastal waters spawn from May to August, and these groups are divided into winter, autumn, and summer populations, respectively.

5.5.3.3 Spawning Habits

In general, mature octopus eggs are produced and stored in the carcass cavity and then expelled from the body cavity. Spawning activity occurs in the substratum in most genera and superficially in a few species (Fig. 5.12). The female behaves as an egg guard and dies when the fertilized eggs hatch. The length of incubation varies from 1 to 3 months, and recent studies have found that *Graneledone boreopacifica* can incubate for up to 53 months. Some other octopuses, such as Bolitaenidae, Amphitretidae, and *Tremoctopus violaceus*, have specific incubation behaviors where fertilized egg output is carried in carapace and inter-carapace membranes until hatching. Ocythoidae and Vitreledonellidae are ovoviviparous, with fertilized eggs incubating in the oviduct.

In *Nautilus*, mature eggs are discharged singly, the substratum spawns, and the eggs are produced and adhere to hard objects. In cuttlefish, mature eggs are produced and stored in the carcass cavity, and eggs are discharged individually. All genera spawn demersally, and fertilized eggs adhere to hard objects on the substrate (Fig. 5.12). Mature eggs of loliginid and ommastrephid are produced and temporarily stored in the oviducts, and at oviposition, the secretory glands of the ovipositor glands encase the eggs in an egg mass and discharge them into the water column (Fig. 5.12).

5.5.3.4 Spawning Strategies

The spawning strategies of cephalopods can be classified into five strategy types based on the type of oviposition, spawning pattern, and individual growth between their first and second spawning events.

Polycyclic Spawning In this spawning strategy, oviposition involves the oocytes maturing and being expelled in batches during spawning; the spawning pattern is multiple rounds of spawning in which the gonads can develop and mature multiple times; the parent continues to survive and grow after spawning and then begins a new round of spawning during the spawning season. This strategy is similar to the multiple reproduction strategy of fish, and *Nautilus* is the only genus of cephalopod to employ such a strategy (Fig. 5.13a).

Simultaneous Terminal Spawning In this spawning strategy, spawning involves both oocytes developing and maturing at once, and all mature eggs are expelled within a short period

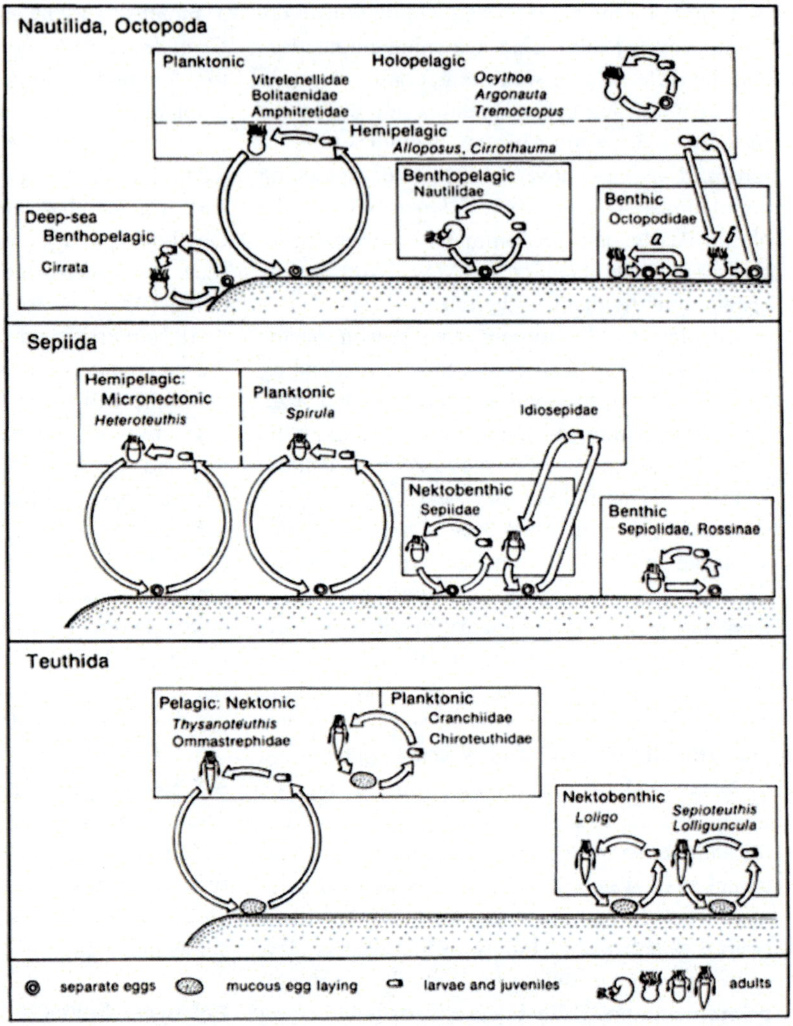

Fig. 5.12 Life history of cephalopods and their egg-laying habits (Arkhipkin 1992)

of time before the end of the parental life; no new oocytes are produced during the spawning cycle; the spawning pattern is a single round of spawning in which the gonads develop and mature at once; individuals stop feeding and growing during spawning and die at the end of spawning (Fig. 5.13b). *Doryteuthis opalescens*, *T. pacificus*, *O. vulgaris*, and *O. mimus* all employ this strategy.

Intermittent Terminal Spawning In this spawning strategy, oviposition involves the single production of oocytes, with both oocytes maturing in batches and mature eggs being produced in batches over a relatively long period of time; the spawning pattern is a single round of spawning, during which individuals stop feeding and growing and die at the end of spawning (Fig. 5.13c). *I. argentinus*, *L. vulgaris*, *L. forbesii*, and *Sepia officinalis* all employ this strategy.

Multiple Spawning In this spawning strategy, oocytes are produced multiple times and mature in batches, and mature eggs are expelled in batches during the breeding season; the spawning pattern is a single round of spawning, with the gonads developing and maturing in one step; individuals continue to feed and grow during

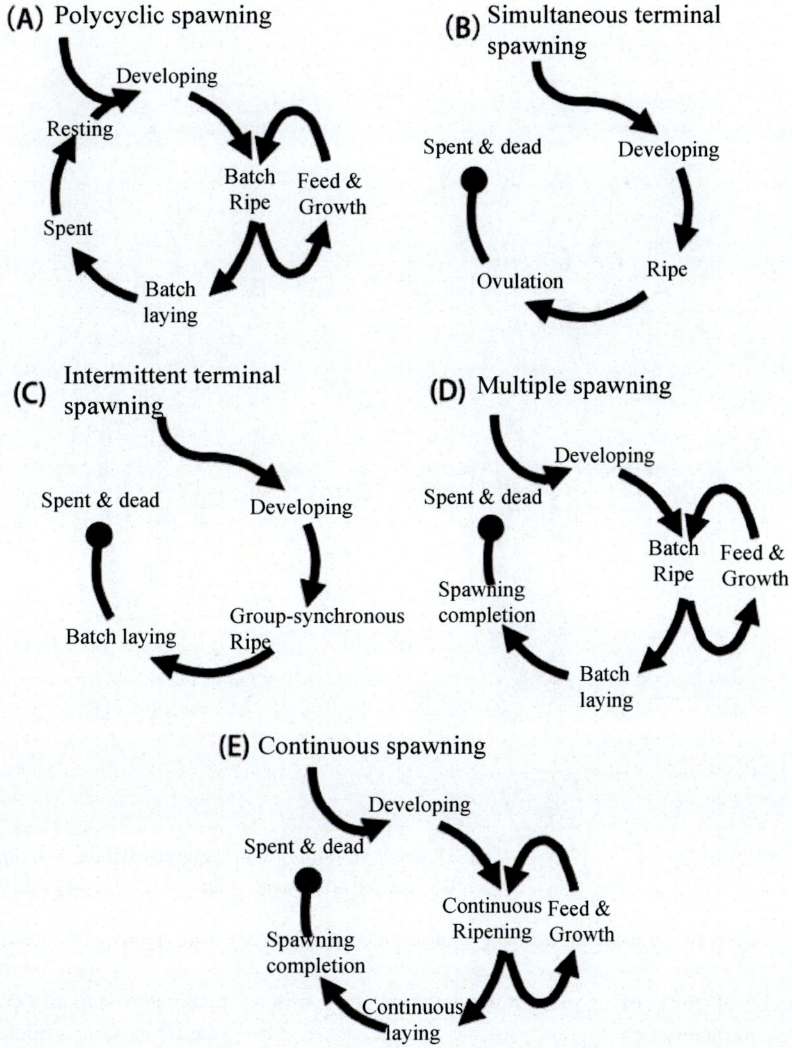

Fig. 5.13 Schematic diagram of cephalopod spawning strategies (Lin 2015)

spawning and die at the end of the breeding season (Fig. 5.13d). *D. gigas*, *S. oualaniensis*, *O. bartramii*, and *N. gouldi* all employ this strategy.

Continuous Spawning In this spawning strategy, oviposition involves a continuous production of oocytes that mature randomly and are expelled; the spawning pattern is a single round of spawning, with the gonads developing and maturing in one step; the individual continues to feed and grow during spawning and dies at the end of the breeding season (Fig. 5.13e). *Cirroteuthis muelleri*, *Opisthoteuthis agassizii*, *Argonauta argo*, *Ocythoe tuberculata*, and species of the genus *Rossia* employ this strategy.

5.5.4 Description of Eggs and Their Maturation

5.5.4.1 Description of Eggs

The eggs of cephalopods are large oocysts containing large amounts of yolk. The eggs are classified according to their specific gravity, the presence or absence of viscosity, and the presence

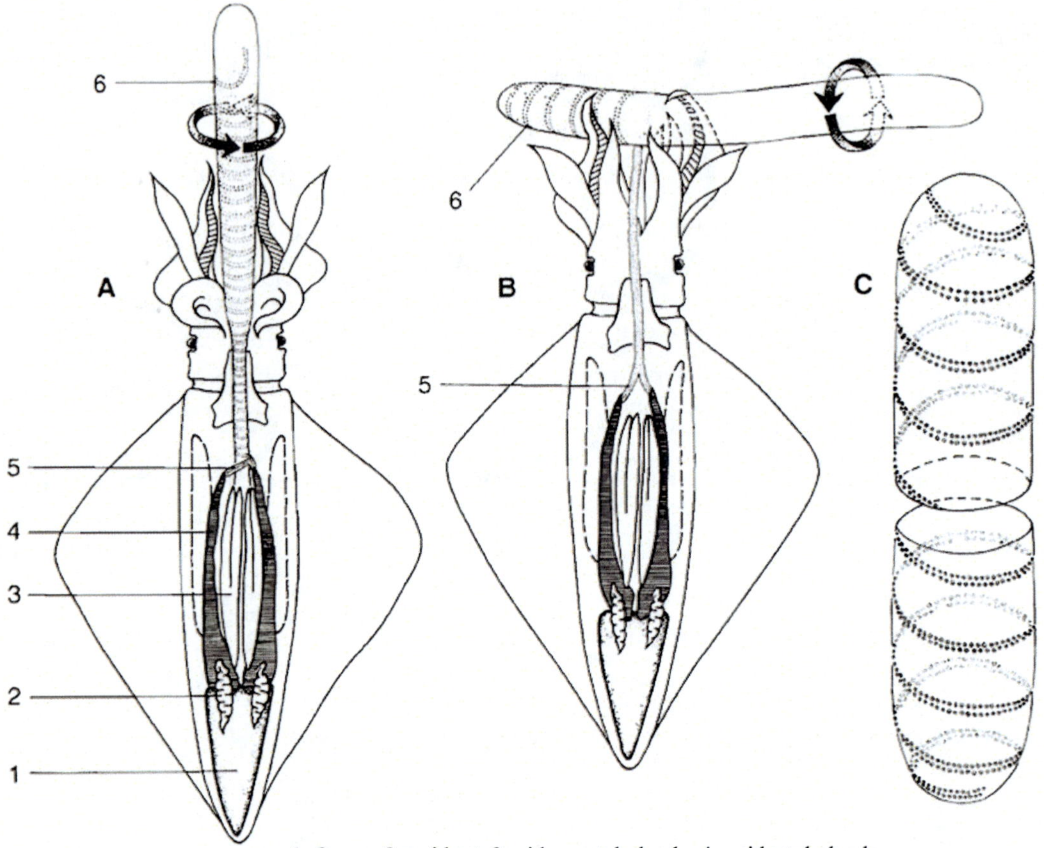

1. Ovary; 2. oviduct; 3. nidamental glands; 4. oviductal glands;
5. secretion of oviductal glands with ripe eggs; 6. egg mass

Fig. 5.14 Scenario of floating egg mass production by the female diamond fin squid (Nigmatullin et al. 1995)

or absence of an oocyst (sheath), among other characteristics.

Floating eggs: these eggs have a specific gravity less than water, and the eggs are produced separated and floating, or they are produced and wrapped around the egg sac (sheath) in egg masses or produced as egg masses and floating. For example, *Watasenia scintillans* and *Vampyroteuthis infernalis* lay separated floating eggs; *Tremoctopus violaceus* wrap egg-attached filaments around egg shafts that float as egg masses; *D. gigas*, *O. pteropus*, and *Thysanoteuthis rhombus* produce eggs that are wrapped in gel secretions by the female that float in egg masses in the water layer (Fig. 5.14).

Sticky eggs: these eggs have a specific gravity greater than that of water, with a sticky or bifurcated stalked oocyte that adheres to aquatic plants, to sand grains, or under rocky reefs after output. Cuttlefish generally lay such sticky eggs, e.g., cuttlefish eggs with a forked stalk at one end, used for attachment to algal corals or branches, aggregating eggs in grape bunches; *Sepia pharaonis* lay cream-colored eggs with thicker egg membranes for attachment to *Sargassum* willow corals or fine branches of sea plants; *Sepia esculenta* eggs have secondary and tertiary egg membranes, used for attachment to the sea floor or large seaweeds, with transparent, slightly cream-colored, subelliptical eggs. *Octopus ocellatus* and *Octopus variabilis* lay eggs attached to empty shells and under rocky reefs, respectively.

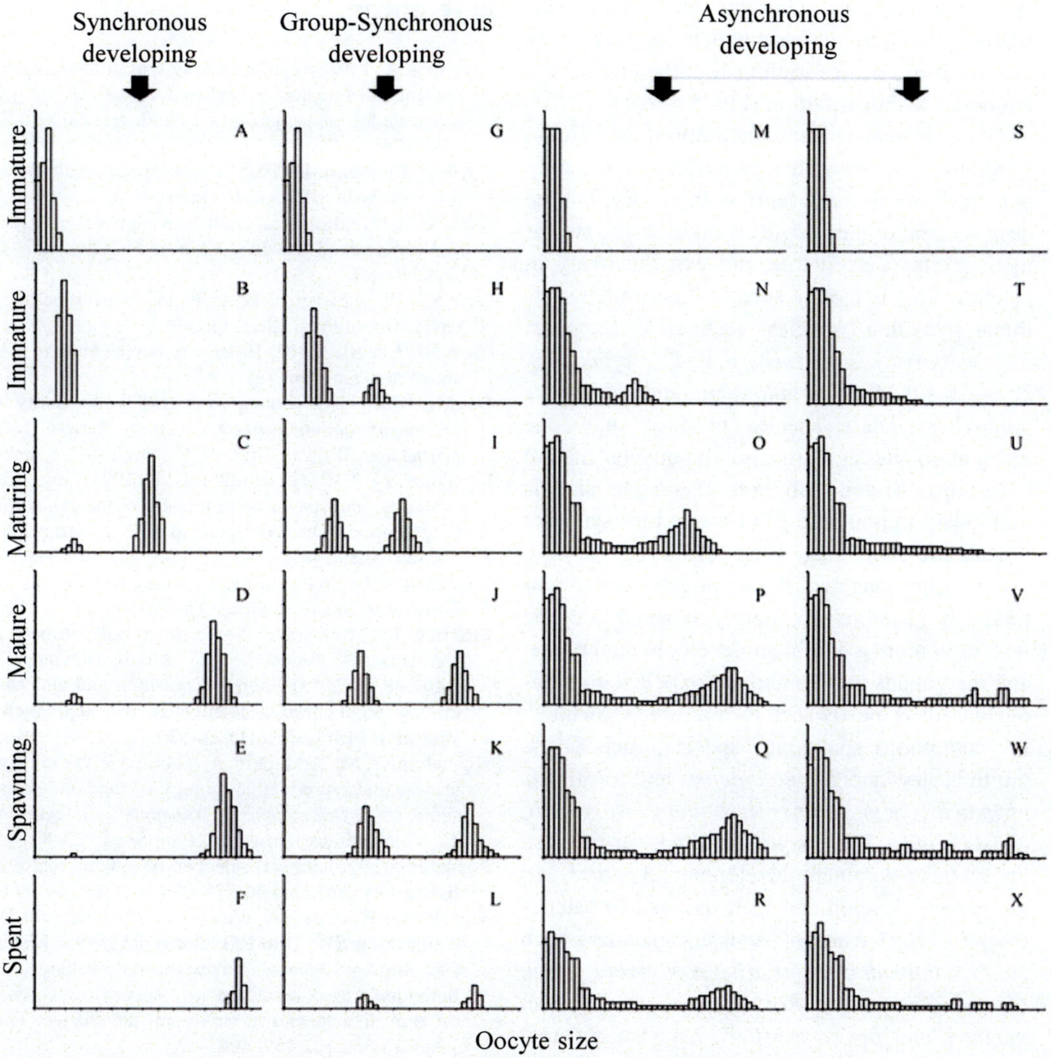

Fig. 5.15 Schematic diagram of cephalopod egg maturation types (Lin 2015)

Sinking eggs: these eggs have a specific gravity greater than water and sink to the bottom after being expelled. For example, Sepiolidae, which inhabits sandy or muddy substrates, lays their eggs on sand grains; eggs of *Idiosepius paradoxus* are produced individually, and the eggs are placed on the bottom of the water body or on the surface of other objects with their antennae after production. In addition, the eggs of *T. pacificus* are also sinkers, but because the female is wrapped in a gelatinous oocyst after spawning, the egg mass is mostly suspended in the water column by currents and often adheres to objects on the bottom.

5.5.4.2 Egg Maturation

Egg maturation in cephalopods can be divided into the following categories:

Single-batch synchronous maturation: the ovary has a single batch of oocytes, all of which are at essentially the same development level, maturing and being expelled from the ovary at once. This type of maturation is mostly seen in "transient terminal spawning" genera, such as

Moroteuthis ingens, *Gonatus antarcticus*, and *Pteroctopus tetracirrhus*, which produce only one batch of oocytes in their lifetime and have a unimodal size distribution (Fig. 5.15a–f).

Grouped synchronous maturation: as with the single-batch synchronous maturation type, only one batch of oocytes occurs in the ovary, but the development of the oocytes is divided into two or more groups that mature and exit the ovary in batches. This is mostly seen in "intermittent terminal spawning" species, such as *T. rhombus*, *L. vulgaris*, *L. forbesii*, *S. officinalis*, *T. violaceus*, *Sepiola atlantica*, and *Scaeurgus unicirrhus*. These species produce only one batch of oocytes at a time, but the oocytes mature in groups. The distribution of oocyte size is unimodal in premature gonads and bimodal after the late immature stage (Fig. 5.15g–l).

Asynchronous maturation: oocytes occur continuously in the ovary, mainly as small oocytes, with or without a dominant developmental phase, and the gonads mature and are expelled continuously. This is mostly seen in "multiple spawning" or "continuous spawning" species, such as the multiple spawners *S. oualaniensis* and *Idiosepius pygmaeus*, *Sepioteuthis lessoniana*, *N. gouldi*, and *D. gigas*. The oocyte development of these genera has a dominant developmental period, and the gonads develop and then oviposit in batches (Fig. 5.15m–r), whereas persistent spawners, such as *Opisthoteuthis agassizii*, *Argonauta argo*, *Ocythoe tuberculata*, *Rossia macrosoma*, *Rossia pacifica*, and *Rossia moelleri*, have no apparent dominant developmental period for oocyte maturation (Fig. 5.15s–x).

References

Arkhipkin A (1992) Reproductive system structure, development and function in cephalopods with a new general scale for maturity stages. J Northwest Atl Fish Sci 12:63–74

Boyle P, Rodhouse P (2005) Cephalopods: ecology and fisheries. Wiley-Blackwell, Oxford

Chen XJ (1991) Biological studies on *Nototodarus gouldi* in New Zealand waters. J Chinese Fish Sci 1:27–32. (In Chinese)

Chen XJ (2014) Fisheries resources and fisheries oceanography. Ocean Press. (In Chinese)

Chen XJ, Liu BL (2017) Fishery resources biology. Science Press. (In Chinese)

Du JR, Chen BQ, Zhang QY (1983) Fecundity of *Trichiurus haumela* in the western Taiwan Strait marine area. Taiwan Strait 01:122–132. (In Chinese)

Laptikhovsky VV, Nigmatullin CM (1993) Egg size, fecundity, and spawning in females of the genus *Illex* (Cephalopoda: Ommastrephidae). ICES J Marine Sci: J Conseil 50:393–403

Lin DM (2015) Study on the spawning strategy of *Illex argentinus*. Shanghai Ocean University. (In Chinese)

McBride RS, Somarakis S, Fitzhugh GR, Albert A, Yaragina NA, Wuenschel MJ, Alonso-Fernández A, Basilone G (2015) Energy acquisition and allocation to egg production in relation to fish reproductive strategies. Fish Fish 16(1):23–57

Nigmatullin CM, Arkhipkin A, Sabirov R (1995) Age, growth and reproductive biology of diamond-shaped squid *Thysanoteuthis rhombus* (Oegopsida: Thysanoteuthidae). Mar Ecol Prog Ser 43:73–87

Pianka ER (1976) Natural selection of optimal reproductive tactics. Am Zool 16:775–784

Van Winkle W, Rose KA, Winemiller KO, Deangelis DL, Christensen SW, Otto RG, Shuter BJ (1993) Linking life history theory, environmental setting, and individual-based modeling to compare responses of different fish species to environmental change. Trans Am Fish Soc 122:459–466

Yellow Sea Fisheries Research Institute (1981) Handbook of marine fisheries resources mobilization (In Chinese)

Fish Prey, Food Habits, and Interspecific Relationships

Xinjun Chen, Bilin Liu, and Yunkai Li

Abstract

The communities of fishery resources and other aquatic organisms are linked by food chains and food webs. The number of a certain group not only is related to the number of its predators and food but also depends to a large extent on the food security of fish; the quantity, quality, and availability of food; the length of feeding season; and the quantity and quality of fish species in the waters which constitute the main contents of the study of feeding habits in fishery resources. Fish feeding ecology is an important part of fish ecology, though feeding ecology research can provide basic data for further analysis and understanding of population dynamics. Based on the analysis of stomach contents, fish feeding ecology can be divided into three levels: individual level, population level, and community level. In recent years, with the development of science and technology and the remarkable progress of element analysis technology, new techniques such as stable isotope method and fatty acid labeling method have been applied in the study of fish feeding ecology. This chapter focuses on the relationship between fish and food chain, the types and characteristics of fish feeding, and the research methods and explains how fish ensure their food supply, and at the same time, the conception and research method of fatness and fat content were also introduced. As a result of the development of marine fishery, the practice of fishery production and management has put forward higher requirements for us, that is, not only the static research but also the dynamic understanding of feeding habits, that is, the changing law with time and space, and the quantitative relationship between predation and predation among populations, these results will provide a mathematical model for resource assessment, in particular, the establishment and improvement of ecosystem-based models for fishery resources assessment and management provide valuable basic information.

Keywords

Fish prey · Food habits · Feeding ecology

Abbreviations

s_i	the standard deviation of the rows
r_i	the proportion of prey organisms in the stomach contents of fish; the percentage of a component in the food

X. Chen (✉) · B. Liu · Y. Li
College of marine sciences, Shanghai Ocean University, Shanghai, China
e-mail: xjchen@shou.edu.cn; Bl-liu@shou.edu.cn; Ykli@shou.edu.cn

X. Chen, B. Liu (eds.), *Biology of Fishery Resources*, https://doi.org/10.1007/978-981-16-6948-4_6

d_{ij}	the Euclidean distance (specimen point distance) between each fish species
I	the selection index
I_i	the selection index of fish for prey i
L	Body length
NGS	next-generation sequencing
p_i	the proportion of the same prey organism as that in the environmental; the percentage of the same component in the food base; Specific prey abundance
Q	fullness
r_i	the proportion of prey organisms in the stomach contents of fish; the percentage of
S_i	the content (volume, weight, and quantity) of prey i in the stomach contents; the standard deviation of the rows
S_{ti}	the stomach content of the ingesting fish with prey i in the stomach
W	body weight
$^{13}C/^{12}C$	the carbon stable isotope ratio
$^{15}N/^{14}N$	the nitrogen stable isotope ratio
$^{18}O/^{16}O$	the oxygen stable isotope ratio
$\delta^{15}N$	nitrogen stable isotope ratios
$\delta^{13}C$	carbon stable isotope ratios

6.1 Feeding Relationships and Food Chains Among Fish

6.1.1 Composition of Fish Feeding

As one of the most important living conditions for fish, prey constitutes the first link in interspecific relationships. The state of fish prey security regulates the growth, development, and reproduction of fish and influences the population dynamics and even the abundance of fisheries.

In general, fish diets are very broad and complex and include aquatic plant groups, ranging from lower unicellular algae to macroalgae and aquatic vascular plants; aquatic animal groups, involving almost every phylum of invertebrates to vertebrate fish; and humic substances, which are also important prey for some bottom-feeding fish.

The composition of prey varies greatly from species to species, with some feeding on plankton in the pelagic zone and others becoming aggressive carnivores feeding on shrimp, crabs, cephalopods, and even their own juveniles, while others prefer organic detritus on the bottom, becoming detritus consumers. Fish species differ greatly among the broad spectrum of their prey. This is the result of the adaptation and evolution of fish over time.

6.1.2 Food Chains, Food Webs, and Their Ecoefficiency

6.1.2.1 Food Chains and Food Webs

The longitudinal and interspecific food relationships among various organisms in aquatic ecosystems, mainly in the form of predation, prey, and competition, form food chains; multiple food chains form a complex weblike structure, collectively known as a food web.

The food chain refers to the food relationship between fish and prey organisms and predators. The relationships are food-primary consumer-subconsumer-higher trophic levels. Several links in a multilevel trophic relationship also have complex intertrophic relationships, such as small animals feeding on smaller animals or plants being preyed upon by larger animals, i.e., lower-level consumers providing food for higher-level consumers. Thus, one link is interlocked with another in a chain-like fashion.

A single animal often feeds on a variety of organisms, and a variety of other organisms are similarly interdependent and nutritionally linked. Therefore, all kinds of organisms in a whole watershed are interconnected and mutually constrained, forming a complex grid-like network called a food web. Food webs are formed gradually during the long-term development of ecosystems and play an important role in maintaining the stability and balance of ecosystems. Each link in a food chain is called a trophic level, which indicates the trophic position of animals in a food web.

The lowest link in the food chain in lakes and oceans is the primary producers, i.e., phytoplankton (unicellular algae and autotrophic bacteria) and macrobenthic algae (including higher vascular plants); the second link involves the animals that feed on the plants, and they are primary consumers, i.e., phytophagous animals (herbivores); then, the third link involves the animals that prey on these animals, i.e., carnivores, and they are secondary consumers, which are secondary predators. This chain continues, and finally, there are heterotrophic bacteria, also known as decomposers (decomposers). They can break down and reduce plant and animal carcasses and debris in lakes and oceans into the nutrient salts needed by the primary producers to grow and reproduce. In this way, nutrient salts are transported through a series of links, forming a closed loop, also known as the food chain. The existence of this food chain is not only a condition for the survival of lakes and marine life but is also an important structure for maintaining the transformation of matter and the flow of energy throughout a watershed. These relationships are of great importance for the development of living resources in lakes or oceans.

Moving from one link to another in the food chain is accompanied by a certain amount of consumption. From an energy point of view, there is a certain conversion rate. For example, phytoplankton are consumed by zooplankton and converted into zooplankton at a projected rate of 20%, zooplankton are consumed by small fish and form small fish at a rate of 10%, and small fish are consumed by large fish at a rate of 10%. This means that to compose one unit of organism of an animal at a higher level in the food chain hierarchy requires approximately ten units of organismal energy of an animal at a lower level in the food hierarchy. It follows that the closer to the first link in the food chain, i.e., the lower level in the food hierarchy, is to the next link, the more abundant the organisms are. This situation is similar to a pyramid, where the number of organisms decreases higher up in the food chain, and this is called the law of the pyramid (Fig. 6.1).

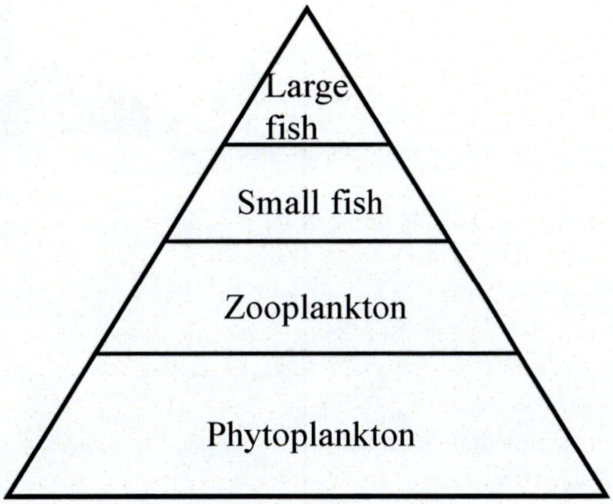

Fig. 6.1 Fish food chain pyramid (Chen 2014; Chen and Liu 2017)

As seen from Fig. 6.2, the interrelationship between the trophic levels of a food web, green plants (producers) are at trophic level 1, phytophagous animals (primary consumers) are at trophic level 2, etc., and from the lower level → higher level in a pyramidal pattern, it is generally believed that the energy conversion efficiency of each trophic level is approximately 10%; that is, when energy is transferred from trophic level to trophic level, only approximately 10% of the energy can be transferred to the next level. The largest share of the world's catch composition is accounted for by pelagic fish. At the lowest level of the food hierarchy, phytoplankton and zooplankton, as well as larval classes, are very large in number and sustain highly productive pelagic fish, which are thus able to support other carnivorous and predatory fish. Therefore, to increase the productivity of waters, it is important to keep the number of rings in the food chain as low as possible; i.e., the closer the caught economic fish is to the first ring in the food chain, the greater the yield obtained. However, the fewer the number of rings in a food chain, the more unstable the ecosystem is and the more vulnerable it is to environmental conditions and other factors.

Most of the fish at lower levels of the food chain are small, low-quality fish, such as Clupeidae, Carangidae, and Scombridae, which are not as high quality as those at higher trophic levels, such as Sciaenidae, Sparidae, Bothidae,

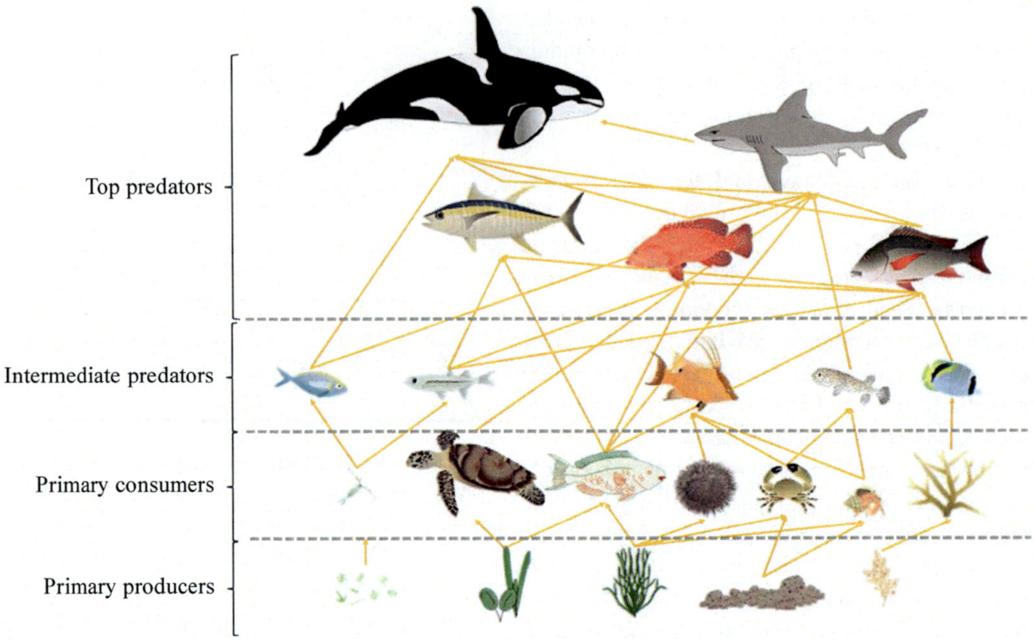

Fig. 6.2 Food webs between catchable marine living resources and primary producers (Chen and Liu 2017)

and Pleuronectidae. Therefore, it is also important to sustainably use fishery resources at different trophic levels.

6.1.2.2 Ecoefficiency

Food chains are pathways through which energy flows in an ecosystem. The transition of energy from one trophic level to another does not remain constant, and each subsequent trophic level of the food chain takes up only a small fraction of the food energy provided by the former trophic level. That is, energy flow is gradually reduced. Lindeman (1942) proposed the "one-tenth" law, which indicates the quantitative relationship between the flow of energy between trophic levels in an ecosystem and the efficiency of energy conversion through different trophic levels, and this law is referred to as Lindeman efficiency or ecoefficiency.

Studies have shown that the amount of energy flow in food chains universally decreases sharply as it passes through trophic levels, but the 10% conversion rate is an estimated average that varies considerably between food chains; typically consumers in an ecosystem can only convert up to 4.5–20% of food energy into their own material (protoplasm). There are usually five trophic levels in a food chain, with green plants as primary producers as the first trophic level, phytophagous animals as the second level, and lower, middle, and higher carnivores as the third, fourth, and fifth levels, respectively. As matter and energy flow from low to high through the trophic levels of the food chain, energy decreases in a stepwise fashion with each trophic level, forming a pyramid shape called the energy cone or ecological pyramid.

6.2 Types of Fish Feeding

Because of the wide range of fish diets, the vast majority of aquatic organisms can be consumed by fish, so the prey of fish is very diverse and extensive. In nature, no fish exists that eats all animals and plants, and it is difficult to find a fish that eats exclusively one individual. Usually, a certain fish can eat dozens or even hundreds of species, and these feeding characteristics are closely related to the chewing organs and foraging methods of fish and are also the result of long-term fish adaptations and evolution. Combined with the feeding characteristics of fish, the feeding types of fish can usually be divided according to the different types of food consumed, food ecological types, predatory nature, and feeding mechanism.

6.2.1 Classification According to the Type of the Food Consumed by a Fish

These fish can usually be classified as phytophagous, zoophagous, and omnivorous.

6.2.1.1 Herbivores

Fish that fed on aquatic plant-based food. These fish can also be divided into four categories according to the nature of their staple food.

1. Fish that feed mainly on phytoplankton. For example, *Konosirus punctatus*, *Sardinops sagax*, and *Hypophthalmichthys molitrix*. The gill rakers of this type of fish are very dense and suitable for filtering planktonic single-celled algae, with well-developed intestinal tubes for nutrient absorption. The number of *K. punctatus* is approximately 285 gill rakers. The length of the intestinal canal is three to eight times the body length.
2. Fish that primarily feed on periphyton. This type of fish has a prominent muzzle that facilitates feeding on filamentous algae attached to a reef, such as *Varicorhinus heratensis* and the large-nosed soft-mouthed fish (*Chondrostoma nasus*).
3. Fish that feed mainly on higher aquatic vascular plants. These fish have strong and well-developed pharyngeal teeth and long intestinal tubes suitable for chewing aquatic. For example, *Ctenopharyngodon idellus* pharyngeal teeth are pectinate, and its basal occipital triangular bone pad for grinding can grind plant stems and leaves and cut them to enhance digestion; its intestinal canal is more than three to eight times its body length, and this fish has high amylase activity.
4. Fish that feed on humic substances and detritus. For example, *Mugil cephalus* has a terminal mouth position and a well-developed muscular stomach, similar to the sand sacs of birds, and these are used to grind single-celled algae. The gill rakers in *Mugil soiuy* are in the range of 61–87, and the intestinal canal is more than three times the body length.

6.2.1.2 Carnivores

Fish that primarily prey on animals are characterized by sparse gill rakers and short intestinal tubes. They can usually be subdivided into the following categories:

1. Fish that feed on zooplankton. These fish include *Clupea pallasii*, *Engraulis japonicus*, *Setipinna gilberti*, *Scomber japonicus*, etc. *C. pallasii* has 63–73 gill rakers and long digestive tubes. They feed on krill, copepods, and Amphipods.
2. Fish that feed on benthic organisms. These fish include Bothidae, Pleuronectidae, and Cynoglossidae that are very prey-rich and have diverse dental morphologies, including pavement, cusp, canine, molar, or rostrum morphologies. The number of gill rakers and the length of the gut tube are intermediate between those of zooplankton feeders and swimmer feeders.
3. Fish that feed on swimming organisms. *Trichiurus haumela*, *Scomberomorus niphonius*, and *Pseudosciaena crocea*, which feed on swimming shrimp and small fish, have sharp teeth, short intestinal tubes, and very high digestive protease activity.

6.2.1.3 Omnivores

An omnivorous fish feeds on plants or animals. These fish are characterized by a medium mouth with conical, narrowly flattened or molar-shaped teeth in both jaws. The gill rakers are medium, the length of the digestive tube is less than that of phytophagous fish, and the amylase and protease enzymes for digesting carbohydrates occur at high levels, facilitating digestive growth.

6.2.2 Classification Based on the Ecological Type of Food Consumed by a Fish

These types of food can be divided into three categories: plankton, swimming organisms, and benthic animals.

1. Fish that feed on plankton. This type of fish is widely distributed and extremely productive, with a predominantly spindle-shaped body, fast swimming speed, strong digestive capacity, and rapid growth in small and medium fish, such as Clupeidae, Engraulidae, and Carangidae.
2. Fish that feed on swimming organisms. This type of fish is large, swims very well, has a large mouth, is very rich in digestive enzymes, grows rapidly, and feeds exclusively on slightly smaller fish, cephalopods, shrimp, and crabs. They have a high fishery value and include *Trichiurus lepturus*, Sciaenidae, and Sparidae.
3. Fish that feed on benthic animals. These fish are sparsely populated and do not form dense schools. Their dentition is highly variable and specialized to suit the diversity of benthic invertebrate types, such as Bothidae, Pleuronectidae, Dasyatidae, Rajidae, and Soleidae.

6.2.3 Classification Based on the Number of Prey Species Consumed

These fish can be divided into generalist and specialist fish.

1. Generalist fish. These fish feed extensively on a variety of prey organisms, and many omnivorous fish belong to this category; for example, *P. crocea* feeds on almost 100 species, and striped bass feeds on 40–60 species of prey.
2. Specialist fish. A small number of fish are distributed in a specific type of water, specializing in hunting certain plants or animal as prey, and their mouthparts and digestive functions are more specialized and cannot easily adapt to the external environmental conditions that can intensely change. Examples of these fish include *Fistularia* spp., *Tylosurus*, *Syngnathus*, and *Hippocampus* spp.

6.2.4 Classification Based on the Nature of Fish Predation

These fish can be divided into two categories: mild fish and aggressive fish.

1. Mild fish. These fish generally feed on small phytoplankton, zooplankton, small benthic invertebrates, organic detritus, or animal carcasses, such as those of *Mugil* spp., *Sphyraenus* spp., *Engraulis* spp., and *Clupea* spp.
2. Aggressive fish. These fish have sharp teeth, are fast swimmers, and live by hunting other fish and smaller invertebrates, and these fish include *Trichiurus lepturus*, *Muraenesox cinereus*, and *Carcharodon carcharias* which can reach a length of nearly 12 m. These fish are extremely aggressive, with sharp triangular teeth and tiny serrated edges, and they can bite large fish or even mammals.

6.2.5 Classification Based on the Way Fish Feed

These fish can be divided into five categories: filter-feeding fish, scraper-feeding fish, predatory fish, sucker fish, and parasitic fish.

1. Filter-feeding fish. Specialized filter-feeding fish are characterized by large mouths, fine gill rakers, and weakly developed teeth, and they take in food directly from the oropharynx into the gastrointestinal digestion. Example species include *Engraulis* spp., *Chanos* spp., etc.
2. Scraper-feeding fish. These fish have unique teeth and oral structures that specialize in scraping organisms from rocks. They have

especially well-developed incisors and include fish such as Tetraodontidae.
3. Predatory fish. Characterized by their swift swimming and sharp teeth, they can quickly and accurately pursue their prey and swallow it in one gulp, and these fish include *Trichiurus lepturus* and *M. cinereus*.
4. Suction-feeding fish. These fish form cylinders with specialized mouths that draw in food and water together, causing an attraction current that draws small plants and animals into their stomachs, and these fish include *Syngnathus* spp. and *Hippocampus* spp.
5. Parasitic fish. These fish feed on the nutrients or excreta of their hosts; e.g., *Carassius auratus* specialize in feeding on the excreta or incompletely digested food of larger fish. For example, a male *Ceratias holboelli* feeds on a female by parasitizing her.

Research has shown that the food habits of fish and other organisms are a product of biological adaptations to the environment. For example, the fact that the environment is more stable at lower latitudes than at higher latitudes makes the prey base of fish at lower latitudes more stable, which results in the diets of high-latitude fish species being generally more extensive than those of low-latitude fish. As another example, certain characteristics of fish digestive systems have been developed during the long evolution of fish, thus determining certain food habits. However, the feeding type of fish and other fishery resource species is not fixed; it has a certain stability but also plasticity, especially for omnivorous fish that have a large active sea area, high mobility, very complex food prey, and therefore increased plasticity.

6.3 Characteristics of Fish Feeding

When and where a fish consumes its food is related to not only its own biology, such as the developmental stages of its life cycle, but also environmental conditions; thus, it is unlikely that a fish will consume the same food at all times throughout its life cycle, and even aggressive fish can only take in small algae and zooplankton during their early life history stages. These are important features of feeding habits that need to be understood.

6.3.1 Different Feeding Habits at Different Developmental Stages

Fish tend to feed on different objects at different stages of their development. This is partly due to changes in their nutritional requirements and feeding organs and partly because their living environment tends to change at different developmental stages. For example, *Channa argus* has different characteristics at different developmental stages, feeding on planktonic crustaceans during the fry stage; switching to shrimp, aquatic insects, and small fish during the juvenile stage; and reaching adulthood with a prey composition almost entirely composed of fish, except for a considerable number of shrimp.

6.3.2 Changes in the Composition of Fish Food Across Life Stages

The food consumed by adult fish varies not only in quantity but also in species composition during different life stages. For example, many fish consume very little food during the reproductive and overwintering periods and a substantial amount during the feeding periods. In the case of *Scomber japonicus*, the composition of its diet during different life stages is as follows:

1. Upper north period (transition from reproduction to feeding). In the area where cold and warm currents meet, its main prey types are copepods, krill, amphipods, Salpidae, juvenile *Engraulis japonicus*, and Myctophidae.
2. Southward phase (feeding period). Its main prey types are hornless giant krill.
3. Overwintering period. Its prey species are mainly warm-water copepods, decapods larvae, Salpidae, amphipods, planktonic mollusks, and *Noctiluca*.

4. Spawning period (reproductive period). Its prey are mainly copepods, amphipods, egg and juvenile of *E. japonicus*, *Doliolum*, and Salpidae.

6.3.3 Changes in the Composition of Fish Food in Different Waters

Fish have evolved over time to adapt to changes in their environment by changing their biological characteristics, and because the composition of prey organisms varies from one water body to another, fish have had to change their food composition to adapt to their environment. In the case of *Katsuwonus pelamis*, its food composition and changes in different waters have been analyzed and are described below:

1. Eastern waters of Tohoku and Hokkaido. Prey mainly consist of *E. japonicus*, *E. japonicus* larvae, squid, and krill.
2. Waters around the Izu Islands. Prey mainly consist of *E. japonicus*, *E. japonicus* larvae, squid, mackerel, krill, and shrimp.
3. Waters around Ogasawara Islands. Prey mainly consist of Exocoetidae, *Katsuwonus pelamis* larvae, squid, Siganidae, and Holocentridae.
4. Waters off the southern coast of Shikoku. Prey mainly consist of *Trachurus* larvae, *Scomber japonicus*, squid, shrimp, and Gonostomatidae.
5. Balintang waters (southern Taiwan province). Prey mainly consist of squid and Carangidae.
6. Tuko-la-Okinawa waters. Prey mainly consist of *Scomber* spp., *Trachurus japonicus* larvae, Exocoetidae, etc.

As mentioned above, *Trachurus japonicus* typically feed on fish, but the variety of their targets varies considerably in different waters, from migratory to sedentary, feeding extensively on plankton and other invertebrates.

6.3.4 Diurnal Variation in Feeding Habits

Due to external environmental conditions such as light, many fish and other prey organisms tend to move vertically, and their feeding habits change diurnally. For example, in one study, the diurnal feeding intensity of black scraper *Thamnaconus modestus* in the East China Sea was greatest from evening to the first half of the night (69.9%), followed by the second half of the night to dawn (27.5%), and the intensity was the lowest in the morning (16.9%); from evening to night, its stomach contents included mainly copepods, isopods, and mesopods of planktonic crustaceans; from the second half of the night to dawn, this fish mainly consumed fish eggs; in addition to mainly consuming fish eggs, the stomach contents of the specimens caught in the morning also included a number of corals.

Numerous observations have revealed that the actions of prey largely determine not only feeding actions but also diurnal variations in food composition.

6.3.5 Fish Prey Selectivity

In general, a fish does not have an equal interest in a large number of prey organisms but has a preference, or fish are selective about their food. However, some species show it more obvious selectivity and some less obvious selectivity. Selectivity can be judged by two factors: the ratio of the values of the various prey organisms in a habitat and the ratio of the number of prey organisms consumed by a fish.

Fish are somewhat plastic in their choice of food. When a fish does not have access to its preferred food, it can still feed on other prey types, especially in the juvenile stage, where plasticity is greater. Depending on their preferences and food availability, the food consumed by fish can usually be divided into the following categories:

1. Main food constitutes the main part and is capable of meeting the needs of life entirely.

2. Secondary food is often seen in the intestines of fish but not in large enough quantities to fully satisfy the needs of a fish.
3. Incidental food is prey that is consumed by fish by chance.

In addition, sometimes, due to changes in environmental conditions, fish lack their main food items and ingest some emergency food. For example, sometimes echinoderms and serpentine animals, which are not normally ingested, can be found in the stomach of fish that are apparently forced to swallow them due to a lack of food.

6.4 Food Security for Fish

6.4.1 Fish Food Security

Fish populations and prey biomass, as well as the total biomass of all fish in a water body, depend to a large extent on food security. Food security means that the waters must not only contain prey organisms that fish can feed on but also have environmental conditions that ensure the possibility for fish to feed. The quantity and quality of prey organisms such as plankton, benthic organisms, and fish, which are used as fish food, are also referred to as the prey base. Thus, food security for fish depends on the quantity, quality, and availability of food in their aquatic habitat.

Fish food security is influenced to some extent by the length of the feeding season, but the length of the season does not always affect fish food security; for example, the vast majority of saltwater bream stop feeding after reaching a certain level of abundance and lipid content and begin their overwinter migration to the sea. In fact, when the prey base is at a low level, the length of the prey season has a limiting effect on fish food security. If the prey base is high, then the effect of the length of the season on fish food security is usually felt only at the margins of their range. Fish food security is also affected by the abiotic environment during the feeding season, such as by changes in temperature, light, wind, waves, size of prey distribution, and many other factors, and to a large extent by the level of defense against predators during the feeding season.

Studies have concluded that fish population size is closely related to biomass and the food security of the species. Fish food security is governed by the following factors: the quantity and quality of prey in the water and its availability, the length of the feeding season, the number of fish seeking prey, biomass, and biomass quality. A fish population influences the prey base, which ensures the growth of that population, the maturity of the sexes, the abundance of the fish, the heterogeneity of individuals within the population, etc. Therefore, when evaluating fish food security, it is best to determine the condition of the fish in terms of growth, abundance, lipid content, heterogeneity of individuals within the population, and other indicators.

6.4.2 Adaptation of Fish to Food Security

Individual fish have evolved adaptations that allow them to make the most of the prey base in their complex environments. Fish from the same community are adapted to consume certain types of prey and, through divergence in their diets, to resolve prey conflicts due to feeding on similar food groups with other species. Adults generally have similar diets only in terms of secondary feeding objects, while the main components of their diets are different; in addition, resolution of feeding conflicts by different feeding periods is less common. In contrast, juvenile fish feeding conflicts are mainly resolved by staggering the time of consumption of similar foods. This is generally much less common in the adult stage because of the different composition of food in the juvenile stage. However, different species of juvenile fish consume certain foods at different times. For example, in the same waters, *Esox* spp., *Lateolabrax* spp., and *Parabramis* spp. juveniles mainly consume similar foods such as rotifers and copepods in their anadromous larval stage; however, these *Lateolabrax* spp. and *Parabramis* spp. reproduce earlier, and by the time the *Lateolabrax* spp. juveniles have switched to feeding on these prey foods, the earlier-born *Esox* juveniles have switched to feeding on other larger size prey. At the same time, the *Lateolabrax* spp. juveniles and *Parabramis* spp. juveniles consume similar foods at longer intervals. Their juveniles are mostly specialists, restricted to feeding on a set species. This scenario may also be the direct cause of the large numbers of mortalities they often suffer, and the prey required for the juveniles is almost always sufficient in water; however, where there is a high concentration of juveniles, there is sometimes not enough prey.

As fish move from one stage to another during development and growth, their food habits also change, an important adaptation for expanding their prey base. During development, in comparison to species with high food security, species with low food security in the early stages of development only switch to external nutrition when they are larger so that they accumulate more yolk in their eggs.

The separation of feeding sites by age is an adaptation that promotes increased food security, and differences in diets between fish of the same size but different sexes are also an adaptive attribute of fish to improve their food security. For example, females of *Sphyrna lewini* often congregate in pelagic waters to feed, while males mainly inhabit nearshore shelf areas; sea cod males consume relatively more crustaceans and worms.

If generations are quite large and food security is low, then a fish population generally shifts to a broad diet, giving it the widest range of recipes. The breadth of a diet varies with food security.

If the parental fish food security is low, then the sizes of its eggs are different, and the duration of hatching small fish from the egg membrane is also different, thus extending the time for juvenile fish to start feeding to the outside world. In the food conversion phase, fish yolk accumulation is different, and the pattern of consuming prey between day and night is also different, thus improving food security. For example, individuals with high yolk accumulation in *Parabramis* spp. will suspend the phenomenon of foraging at night, while individuals with low yolk accumulation will forage all day. Most of the eggs produced by females with low food security produce different sizes of hatchlings, so the level of prey available to the hatchlings in the same period varies in the environment. Even if the eggs hatch at the same time, the prey base expands slightly when they are transferred to external nutrition, with smaller individuals consuming some types of prey and larger individuals feeding on others. After a certain decrease in food security, fish that were previously the same size began to have different growth rates, with some individuals beginning to grow faster and most lagging behind. Fast-growing individuals move to the next developmental stage earlier, e.g., *Carassius auratus gibelio* in the North Kazakh Lake. Slow-growing individuals have a simpler diet, feeding mainly on detritus, while fast-growing individuals feed mainly on shaker larvae, thus increasing food security.

Migration is an important adaptation for fish to improve their food security, and the sizes of their range change as the density of fish populations change. When fish populations are reduced for one reason or another, they sometimes significantly reduce the size of their feeding grounds. For example, *Clupea harengus* and *Gadus morhua* have reduced feeding ranges and change distances as their numbers decrease.

The clustered lifestyle of many fish species during feeding is an important adaptive attribute for maximum food security. This is evident in pelagic fish, where clustering allows them to locate prey directly and easily and facilitates protection against predators and migration. In comparison to individual fish, groups of fish are more likely to locate and remain in contact with dense groups of prey, and in comparison to a group, an individual fish is more likely to miss a dense, moving group of prey. Certain aggressive fish form schools, making it easier to locate and maintain contact with moving prey and facilitate direct predation. The feeding activity of fish in schools is more intense than that of fish in a dispersed state. Fish in schools generally feed and digest at similar rhythms, which allows the feeding activity of the entire school to begin and end at the same time, making dense groups of prey easily accessible to the fish. Thus, forming schools during feeding allows fish to expend less energy searching for prey, which means that their food security is improved.

Many fish ingest their own eggs and juveniles when populations experience abundant generations and older fish have unstable food security. This activity is an important adaptive approach for expanding the prey base and regulating fish numbers to match the available prey base in the water, and this activity occurs in *Gadus* spp., *Scomber* spp., *Osmerus* spp., *Esox* spp., *Lateolabrax* spp., and many other fish species.

6.4.3 Impact of the Physical and Chemical Environments of a Watershed on Food Security

Changes in the physical and chemical environmental conditions of a watershed greatly affect the food security of fish.

1. Water temperature. For example, the more days in a year that the water temperature is above 14 °C in a lake in England, the faster the growth of *Micropterus salmoides* that year because the appropriate temperature encourages the growth and reproduction of prey organisms, thus increasing the abundance of prey and promoting an increase in fish metabolism. As a result, fish are able to grow faster. Conversely, if the water temperature is lower than normal for the year, it can reduce the metabolic rate of fish, causing them to grow more slowly. For example, there is a clear relationship between the hydrological condition of carp during their feeding season and the fat content and weight of sex products in their liver.
2. Light. The length and intensity of light have an effect on the feeding activity of fish, especially fish that use visual discrimination for food identification, and light is more significant in their foraging process. For example, river cod is likely to feed on *Leucaspius delineatus* when the light is higher than 0.11 x.
3. Waves. Shallow seas are only 8–10 m deep. When storms hit, affecting shallow seas and causing large waves, the waves hit from the bottom to the surface. Some organisms that feed on benthic fish, such as *Abramis brama orientalis*, stop feeding and immediately come up to the surface.
4. Wind. Powerful winds can affect the distribution of insects on land, such as in England in May, August, and September each year in the windy season, and insects can be affected by wind blowing so that the creek, pond, and lake prey increase, making the growth and development of freshwater salmon very favorable. In the Zhoushan area of China, insects on land are pushed into the shallow sea area every autumn due to the influence of wind so that the number of insects in this sea area increases sharply, increasing the prey for fish, especially supplementing those in the juvenile stage.
5. Sea currents. Seawater influences the distribution of prey; for example, the population size of *Engraulis ringens* is closely related to the distribution and abundance of plankton. If tropical warm currents enter Peruvian offshore fishing grounds, thus leading to a decline in the amount of plankton in the fishery, then the anchovy catch is reduced. For example, the annual production of the Peruvian anchovy in 1970 reached 13 million t, and it was mainly distributed off Peru, feeding on abundant plankton. Growth and development was very rapid. In 1972, due to the El Niño phenomenon, plankton was affected by the decline in the productivity of the waters, and the spawning rate of the Peruvian anchovy was also greatly reduced, only one-seventh of the usual, resulting in a significant decline in catches (3.319 million t in 1975, declining to 823,000 t in 1980).
6. Substrate. The distribution and abundance of benthic animals varies with the substrate and affects the amount of energy consumed during fish foraging. There is a close relationship between the energy consumed by foraging

fish and the level of metabolism. For example, sandy, muddy, rocky, and deep-sea areas are different and inhabited by different prey organisms, and the energy consumed by fish foraging activities is naturally different.

Thus, in comparison to other factors, external abiotic environmental conditions are of greater significance for food security. However, these conditions do not affect food security in isolation but work in conjunction with biological conditions.

6.5 Methods of Studying Fish Feeding

The standard method for studying fish feeding in modern fish ecology is gastric content analysis. Its purpose is to estimate the trophic structure of the fish community and the trophic level of each fish in the community and further study the cycling of material in the food chain and food web of the ecosystem. Visual, frequency of occurrence, counting, volumetric, and weight methods are the main traditional methods used in fish feeding studies. In recent years, with the development of science and technology and remarkable advances in elemental analysis techniques, new technical tools such as stable isotope tracing, characteristic fatty acid labeling, and DNA barcoding have been continuously applied in the study of fish feeding ecology (Chen 2014; Chen and Liu 2017).

6.5.1 Sample Collection and Processing

6.5.1.1 Sample Collection

Due to the degradable nature of animal proteins, fish gut samples must be strictly standardized to ensure reliable analytical results. Thus, the following must be achieved in conducting sample collection:

1. Samples should be fresh or recently have died. When the fish are caught, samples should be taken immediately to avoid affecting the accuracy of the analysis by continuing enzymatic digestion of the stomach contents over time.
2. Samples should be highly representative. Samples that are truly representative of the target group under study. In the analysis of gastrointestinal contents of fish, samples of all sizes should be taken. In terms of fishing tools, set nets, fish cages, and longlines are generally less representative and are only available for reference, while samples taken from trawls, seines, and drift nets are more representative and can be used for analysis. Gastrointestinal samples caught by tools such as set nets or fish cages have been held for an extended period, and most of the food in the gut has been digested or excreted, which seriously affects the accuracy of the gut analysis; in addition, samples caught by tools such as longlines have a high rate of empty stomachs. Sampling large numbers of fish with mobile gear such as trawls, seines, and drift nets is generally preferred.
3. A certain number of samples should be collected. In fishery resource survey studies, the number of samples taken is usually 1/4–1/8 of the total number of catches, in units of one sample per net. The samples are then numbered, placed in bags, and fixed with 5–8% formalin solution. The length, weight, sex, and gonadal maturity of the fish are recorded during the gastrointestinal analysis for control purposes.

6.5.1.2 Treatment of Stomach Contents

Fish stomach contents must be handled carefully and delicately. Because of the strong digestive capacity of fish, it is important to analyze the stomach contents in a timely manner in a predigested or undigested state so that the analysis of the type and quantity of prey can be carried out accurately.

Before identifying the stomach contents, the number of species, weight, and other parameters of the prey types in the sample area need to be determined. The prey in the stomachs of carnivorous fish can be identified based on the shape and size of scales, otoliths, hyomandibular bone, cleithrum, operculum, pharyngeal teeth, and fin rays. They prey in the stomach of herbivorous fish and fish feeding on zooplankton can be identified based on the size and number of stems, leaves, fruits, seeds, shape of zooplankton, appendages, mouthparts, bristles, etc. of aquatic plants in the stomach. The identification of prey can occur gradually in the stomach from shallow to deep areas and should not be done roughly.

6.5.2 Field Observations of Fish Feeding

Since fishery resource surveys often need to be conducted in relatively difficult environments where highly precise analytical results cannot be obtained, a good approach is to use simple, easy-to-use research methods, such as the visual method. The so-called visual method is to judge the proportional size of each prey item to the volume of the entire stomach by directly estimating the volume of each prey item as a percentage of the volume of the entire stomach contents determined with the naked eye.

When the stomach contents of fish cannot be determined by examining the stomach sac alone, other parts of the digestive organs, such as the intestine, may be used. The visual method is discussed in detail by Soviet scholars in the Methodological Guide to the Study of Fish Feeding, which is described as follows:

1. Cylopob E.K. Classifies feeding classes as follows:
   Level 00: no food either in the stomach or in the intestines.
   Level 0: no food in the stomach but residual food in the intestines.
   Level 1: small amount of food in the stomach.
   Level 2: moderate amount of food in the stomach or 1/2 of the stomach.
   Level 3: stomach filled with food, but the lining of the stomach does not expand.
   Level 4: stomach is full of food, and the lining of the stomach is distended.
2. Eotopob T.B. classifies plankton-feeding fish as follows:
   Level A: gastric enlargement.
   Level B: full stomach.
   Level C: moderately full.
   Level D: a small amount of food.
   Level E: empty stomach.
3. Eotopob T.B. classifies benthic-feeding fish as follows:
   Level 0: empty stomach.
   Level 1: very few food items.
   Level 2: a small amount of food items.
   Level 3: multiple quantities of food items.
   Level 4: very large quantities of food.

6.5.3 Qualitative and Quantitative Versus Analytical Methods

6.5.3.1 Qualitative Analysis Methods

Samples must be thoroughly prepared to ensure a qualitative analysis of prey is easy to conduct. For a qualitative analysis, it is best to take food pieces from the stomach and the front of the intestines, as the prey is more intact and easier to identify there, and if the food has begun to be digested, then it will need to be identified on the basis of residue.

Large prey can be identified by the naked eye, while small prey can be identified with the aid of a dissecting microscope. Depending on the requirements of the biological investigation, identification can be carried out to the phylum, order, family, genus, or even the species level. The degree of digestibility of the prey can be determined by comparing the anterior and posterior segments of the digestive tract.

6.5.3.2 Quantitative Analysis Methods

Counting Method

For the counting method, also known as the individual method, the number of individuals of each prey organism consumed by a fish is counted separately in terms of the number of individuals, and then, the percentage of each prey organism of the total number of individuals is calculated. That is, the number of individuals of a particular type (class) of food component in the stomach contents is calculated as a percentage of the total number of individuals of the food component in the stomach contents. For example, if the stomach of a *Decapterus maruadsi* contains 100 copepods, 75 krill, 50 mysids, 20 Macrura, and 5 ostracods, then the percentage of each type of prey type in terms of the total number of individuals is 40%, 30%, 20%, 8%, and 2%, respectively.

The method is rapid, simple, and practical if the food composition can be easily determined. It is particularly convenient in certain situations, such as the analysis of stomach contents of fish-feeding and plankton-feeding fish with similar individual food sizes, where although plankton counts are cumbersome, they can be simplified with the aid of auxiliary sampling, i.e., partial sampling from a known volume of homogeneous water, counting the number of microscopic organisms, and then calculating the total number of organisms, which can be conducted with the aid of Sedgewick Rafter counting baskets (Chen and Liu 2017).

The individual method also does not provide a complete picture of the composition of consumed food alone and is limited by the following factors: (1) the individual method overemphasizes the importance of small organisms that are consumed in large numbers; however, in some cases, small organisms may be overlooked in the food composition because they are digested rapidly; (2) it is difficult to count the number of individuals of all food components because many organisms such as protozoa become paste-like before reaching the stomach sac; (3) the effect of fish size is not taken into account; (4) this method is not applicable to combined foods such as macroalgae and detritus; and (5) the individual prey components derived from this method are often miscalculated because the individual sizes and nutritional values of the various prey organisms are inconsistent, and it is unreasonable to view them in equal amounts. This method is usually used in conjunction with other methods, especially the weight method.

Weight Method

There are two methods used to determine weight: the first method involves cutting open the stomach of a fish, and removing the contents and immediately weighing them; the second method is the corrected weight method, and it involves carefully selecting intact individual prey organisms during any time of the prey analysis, measuring their lengths or estimating their sizes and weighting them so that after a period of time, the range of sizes and weights of various prey organisms can be known and then multiplied by the number of individual weights. Percent food weight refers to the corrected weight of a particular food as a percentage of the corrected weight of the food mass. There are two types of stomach weights: dry weight and wet weight. The wet weight is generally easier to measure; the dry weight is more time consuming, but it is needed for calculating the energy balance. The corrected weight can be used to calculate the percentage of a prey component to the total weight of the stomach contents according to the following formula:

$$\textit{Percent by weight} = \frac{\textit{Corrected weight of particular food}}{\textit{Corrected weight of food mass}} \times 100$$

In addition to weight percentage, the fullness index can also help analyze the weight of the stomach contents. The total fullness index is defined as the immediate weight of the stomach contents multiplied by 10,000 divided by the net weight of the fish, and the resulting value of the 10,000 parts per million ratio can be expressed by the following formula:

$$\textit{Total fullness index} = \frac{\textit{immediate weight of the stomach contents}}{\textit{net weight of fish}} \times 10000$$

If the weight of the food mass used in the calculation is the corrected weight, then the actual weight of the food mass in the above equation is changed to the corrected weight, and the resulting figure is the corrected total index of fullness. The corrected total index of fullness is more correct than the total index of fullness derived from the immediately obtained weight. If the components of the stomach contents are separated and the immediate weight of each component is multiplied by 10,000 and divided by the net weight, the value of the 10,000 parts per million ratio is called the fullness index of that component; similarly, the corrected fullness index of that component can be obtained. Again, the corrected fullness index is more correct than the fullness index derived from the immediate weight. The weight stated above is actually the wet weight of the prey.

When determining the wet weight, the water in a food item is usually drained by filter paper, but moisture is still an important cause of error and needs to be further reduced by letting it dry naturally, drying it on a hot plate, or centrifuging it. The determination of dry weight can be conducted by evaporating the water from the food item to a constant weight, and the dry weight temperature varies for different food types (generally between 60 °C and 150 °C). A temperature that is too high may lead to the loss of volatile fat, and the process is time consuming; thus, freeze-drying is more effective.

In food importance studies, the weight method tends to overestimate the importance of individual large food components, and in addition, the weight of food soaked in formalin is different from its wet weight in the field, with resulting measurement errors. This method is somewhat less applicable than the volumetric method but is largely applicable to the analysis of biological components of typical prey items.

Volume Method

Fish food volume composition refers to the volume of a particular type (class) of food as a percentage of the total volume of the stomach contents. The total volume or fractional volume of the stomach contents is generally determined by the drainage method to determine the percentage of volume of each type of food. A small, graduated test tube or centrifuge tube is commonly used and is filled with 5–10 ml of water. Then, the food mass is placed on filter paper, blotted dry, and left until moist. The food mass is then placed in a known graduated test tube, and the total volume at this point is used to determine exactly how much water has drained. The composition of the food mass for each major type of prey food is calculated as a percentage of the frequency of occurrence or as a percentage of the number of individuals.

This method is more complicated, and the analysis is cumbersome; however, it allows the volume to be determined more accurately, and then, the weight can be determined. Few people use this method because of its duration.

Frequency of Occurrence Method

This is one of the simplest and most commonly used methods for determining the composition of prey. The frequency of occurrence is the number of stomachs containing a particular food component as a percentage of the total number of stomachs (nonempty stomachs). Its specific formula is the following:

$$\textit{Frequency of occurrence} = \frac{\textit{The number of stomachs containing a particular food}}{\textit{The total number of stomachs}} \times 100$$

The frequency of occurrence method has the advantage of being rapid and requiring less instrumentation, but it does not express the relative quantity or volume of each type of prey in the stomach. Nevertheless, this method can provide a qualitative analysis of the types of prey consumed by a fish.

This method is quick and easy if the type of food is easily identified, but it provides only a rough picture of one aspect of fish diets, i.e., the degree of fish preference for a particular food; it does not provide a clear indication of the proportion of a particular (type of) food component to the quantity and volume of the stomach contents.

The advantages of the above methods, such as frequency of occurrence, F%; percent quantity, N %; percent volume, V%; and percent weight, W %, are the comparability of the methods in evaluating the importance of each prey category and their ease in terms of access and handling, where the percent weight indicator can be expressed as wet weight, dry weight, or corrected percent weight. Each method has advantages when evaluating fish feeding habits. The frequency of occurrence method reflects the preference of a fish for a particular prey organism, and the percentage of the quantity provides a good indication of the food composition in the stomachs of fish with similar individual food sizes. Since the weight (volume) of food is related to its caloric value, the weight percentage (volume percentage) reflects the proportion of the total consumption of each prey category by a population. However, these metrics also have some limitations. The frequency of occurrence does not accurately express the actual proportion of each prey organism in a stomach. Percent quantity does not objectively describe the feeding habits of fish whose food varies greatly in terms of individual size, and both frequency of occurrence and percent quantity are strongly influenced by small prey species. Total weight (volume) percentages overemphasize the importance of individual predation on the portion of food that exceeds that utilized by a fish population.

Integrated Graphical Method

In general, data results are more easily understood as graphical representations. The graphical method uses the frequency and relative abundance of prey as coordinates to directly describe prey composition, the relative importance of prey (primary or incidental food), and the evenness of food selection among predators. This method uses a two-dimensional plot of specific prey abundance and prey frequency to show the importance of the prey, the feeding strategy of the predator, and the composition of the ecological niche width and interindividual composition (Fig. 6.3).

The modified Costello plotting method uses specific prey abundance and frequency of occurrence as indicators to form a two-dimensional plot (Fig. 6.4a). Specific prey abundance, P_i, and frequency of occurrence are expressed as fractions:

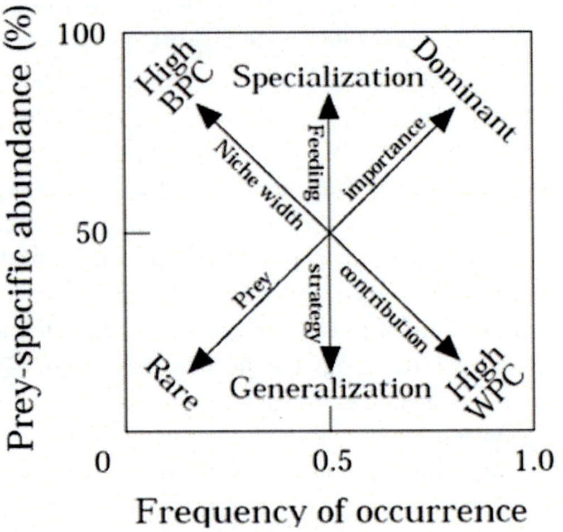

Fig. 6.3 Costello's improved method for interpretation of feeding strategy, trophic niche width contribution, and prey (Amundsen et al. 1996)

$$P_i = \left(\frac{\sum S_i}{\sum S_{ti}}\right) \times 100$$

where S_i is the content (volume, weight, and quantity) of prey i in the stomach contents and S_{ti} is the stomach content of the ingesting fish with prey i in the stomach.

The product of a particular prey abundance and frequency of occurrence corresponds to prey abundance and can be represented by a box enclosed together with the coordinate axes (Fig. 6.4b). The sum of the box areas for all prey species equals the total area of the plot (100% abundance), the product of any particular prey abundance and frequency of occurrence represents a particular prey abundance, and different values of prey abundance can be represented by contours in the plot (Fig. 6.4c).

Using Costello's modified graphical method, information about the feeding strategy of a predator and the importance of prey can be inferred by looking at the scatter of points distributed along the diagonal and axes (Fig. 6.3). On the vertical axis, the feeding strategy of a predator is elucidated according to whether it is broadly or narrowly feeding, and the width of the ecological niche of the predator population can be discerned

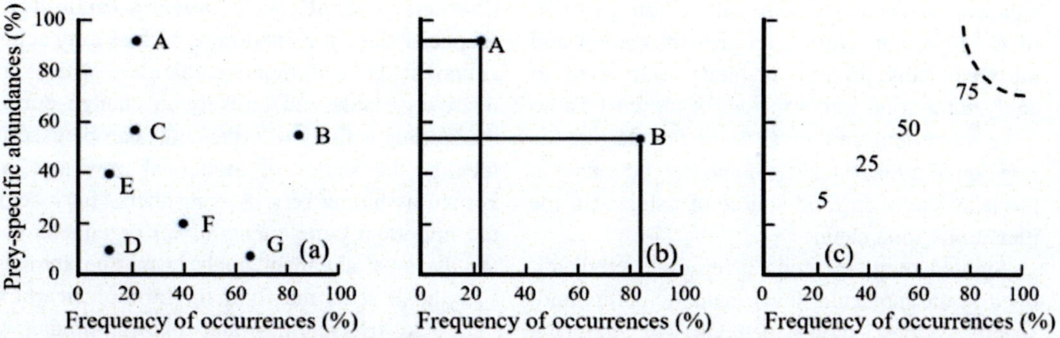

Fig. 6.4 Feeding strategy (Yang and Xie 2000) (**a**). Hypothetical examples (A, B, C, etc. are different prey types). (**b**). Prey-specific abundances of prey types A and B indicated by enclosed areas. (**c**). Isolines representing different values of prey abundances

by observing the position of the values in the plot. The percentage increase in abundance along the diagonal from the bottom left to the top right is used as a measure of prey importance, with important prey (primary food) at the top and nonimportant prey (secondary food or incidental) at the bottom.

The advantage of the graphical method is that the data can be compared quickly and visually on a graph before further statistical analysis. Rather than simply adding or multiplying weight percentages and frequency of occurrence, the graphical method allows a more detailed fractionation of results by weight percentages and frequency of occurrence to distinguish large prey that are present in only a few fish from small prey that are present in many fish, allowing for a better comparison of results.

Stable Isotope Tracing

The stable isotope method is a new biogeochemical research method that has emerged in recent years, and it has been widely applied in the study of aquatic ecosystems and has gradually become a powerful tool for fish feeding analysis, trophic level determination, and food web research. Stable isotopes are different stable forms of the same element with different tissue numbers of the same proton number (e.g., carbon has stable isotopes ^{12}C and ^{13}C). The natural abundance of stable isotopes varies in the environment, and because of their complex fractionation mechanisms during biological metabolism (heavier isotopes linger and become enriched), stable isotope ratios in organisms can be used to trace the flow of material through ecosystems and can provide longer-term information on the feeding of organisms and the transfer of material and energy in food webs. For example, the carbon stable isotope ratio ($^{13}C/^{12}C$) is less enriched between trophic levels in the food chain, at approximately 0–1‰, so it is used to indicate food sources and analyze feeding transformation; the nitrogen stable isotope ratio ($^{15}N/^{14}N$) is more enriched between trophic levels, at approximately 3‰, and is used for fish trophic level determination; the oxygen stable isotope ratio ($^{18}O/^{16}O$) in fish otoliths and the water temperature of habitats have been found to have a linear relationship and are mostly used for life history and habitat reconstruction.

Compared with traditional gastric analysis methods, stable isotope methods provide a more rapid and objective approach for analyzing food web structure and describing the position of food webs in terms of the long-term and short-term changes in organisms' food habits and trophic flows. Stable isotope methods have been widely applied to study energy flow between trophic levels of food chains in lakes, oceans, and estuaries and to map food chain and trophic flow relationships in these ecosystems. For example, carbon and nitrogen stable isotope methods were used to analyze the feeding habits of *Erisphex potti* in the estuary of the Yangtze River and its adjacent waters and to verify the feeding pattern of *Erisphex potti* during its growth; carbon and nitrogen stable isotope methods were used to analyze the food web structure of the East Taihu Lake ecosystem, and it was found that despite the large quantity and quality of submerged plants in the lake, the main food source of fish is still the planktonic food chain.

Stable isotope methods have now been developed to the molecular level, namely, compound-specific stable isotope methods. By analyzing stable isotope information (mainly elemental carbon and nitrogen) at the level of specific macromolecules (e.g., amino acids and fatty acids) in organisms, the patterns of material and energy transfer processes of isotopic fractionation in biological tissues can be determined more specifically. In studies of fish feeding ecology, the analysis of $\delta^{15}N$ and $\delta^{13}C$ in specific amino acids in fish is the main focus. McClelland and Montoya (2002) studied nitrogen stable isotope ratios ($\delta^{15}N$) in marine phytoplankton and showed that changes in $\delta^{15}N$ values in fish with increasing trophic level are the average result of changes in $\delta^{15}N$ values of amino acids in their bodies. Different mechanisms of nitrogen fractionation exist for different amino acids during synthesis and metabolism, where ^{15}N on glutamate is subject to deamination during metabolism and therefore has a high enrichment between trophic levels, up to 7‰ on average, while phenylalanine has an enrichment close to 0 between trophic levels. Therefore, $\delta^{15}N$ values for glutamate can be used for trophic level calculations, while $\delta^{15}N$ on phenylalanine can be used to indicate food sources. Dale et al. (2011) applied $\delta^{15}N$ to glutamate to estimate trophic levels in the *Lutjanus lutjanus* and found that the $\delta^{15}N$ discriminant for glutamate in this fish was approximately 5‰. Notably, both glutamate and phenylalanine are essential amino acids for fish, and essential amino acids cannot be synthesized by the fish themselves and must be obtained from food; therefore, the enrichment of $\delta^{13}C$ in essential amino acids is 0, which makes $\delta^{13}C$ in essential amino acids a more accurate indicator of food source.

Characteristic Fatty Acid Labeling Technology

A special class of compounds, such as fatty acids, amino acids, and monosaccharides, which are relatively stable and unlikely to change during the feeding activity of organisms, can be used to identify the source of biological prey and are known as biomarkers. Among them, fatty acids are important components of all organisms and are the most abundant lipids in marine animals. Fatty acids occur mainly in the form of tricarboxylic acid triglycerides and phospholipids. As biomarkers, fatty acids have several advantages: first, the composition and accumulation of fatty acids in organisms are the result of long-term feeding activities, and there is little chance of incorrectly determining organismal food habits on the basis of fatty acids; second, fatty acids are relatively stable during the metabolism of organisms, and their structure remains essentially unchanged after digestion and absorption by organisms; and third, fatty acids in tricarboxylic acid triglycerides in organisms mainly come from the food consumed, and the use of such fatty acids as biomarkers is generally accepted both nationally and abroad. The composition of fatty acids in an organism is directly related to its food intake.

The use of fatty acids as molecular markers in ecosystems has developed rapidly in recent decades. By comparing the composition of fatty acids among organisms, it is possible to trace the process of material transfer in food webs, to indicate the source of organic matter in food webs, and to contribute to the determination of trophic relationships among organisms. Similar to the results reflected by carbon and nitrogen isotopes, the fatty acid composition of organisms is the result of their long-term feeding activities. Phytoplankton-based autotrophs in marine ecosystems can synthesize all the fatty acids they need, and each phylum of microalgae has its own distinctive fatty acid composition characteristics. For example, the major fatty acids of Pyrrophyta are 16:0, 18:4ω3, 18:5ω3, 20:3ω6, and 22:6ω3; the major fatty acids of Bacillariophyta are 14:0, 16:0, 16:1ω7, and 20:5ω3; and the major fatty acids of Chlorophyta are 16:0, 16:4ω3, and 18:3ω3. In addition, 18:4ω3 in Pyrrophyta and 16:1ω7 in Bacillariophyta have been used as characteristic diatom fatty acids to indicate the methanogenic and diatomic components of particulate suspensions in naturally occurring waters. In addition, phytophagous copepods synthesize large amounts of eicosanoid and docosanoid fatty acids and fatty alcohols, both of which are synthesized by copepods from 18:1ω9 through carbon chain extension. In contrast, omnivorous zooplankton and fish synthesize only small amounts of saturated fatty acids, and the proportion of fatty acid synthesis in the diet of predators decreases at higher trophic levels.

Pethybridge et al. (2010) used multivariate analysis of characteristic fatty acids as a complement to gastric content analysis to determine the fatty acid composition of muscle, liver, and digestive glands in 16 cartilaginous fish species collected from the continental slope of Tasmania, Australia. Gastric and fatty acid composition analyses revealed that each species has a specific feeding and survival pattern (e.g., Chimaeridae are low trophic level organisms in this area, feeding mainly on benthic organisms; medium-sized Squalidae mainly feed on fish and cephalopods in the "middle" of the trophic level; and Scyliorhinidae mainly feed on cephalopods). The compositions of the diets of sharks inhabiting different pelagic levels vary considerably; the diets of different species at the same pelagic level are somewhat similar, and there is a divergence in the ecological niches of feeding; in addition, the fatty acid compositions of different tissues vary, but combining the fatty acid composition of muscle, liver, and digestive glands provides continuous and comprehensive information on the diet of an organism over time.

DNA Molecular Technology

DNA barcoding is a technique that has emerged in the twenty-first century that enables rapid and accurate species identification based on a short, common standard DNA sequence or segments, and this technique has been widely used in the fields of marine fish classification, interspecific kinship identification, molecular genetic diversity, gut microbial diversity, feeding analysis, etc. DNA barcoding is based on the uniqueness of biological genetic sequences in nature. Aguilar et al. (2017) used DNA barcoding to identify the stomach contents of a native catfish and two invasive catfish species in the Chesapeake Bay, USA, and successfully identified 92% of the species, including *Morone americana* and *Anguilla rostrata*, to the species level. The prey abundance of juvenile *Oncorhynchus keta* was comparatively analyzed using morphological observations and DNA barcoding techniques (Sakaguchi et al. 2017). The results showed that 11 of the 36 food species observed based on morphology failed to be detected by the DNA barcoding technique in an analysis of the stomach contents of juvenile fish, while 61 of 80 food species detected by the DNA barcoding technique could not be discriminated by morphology. Thus, DNA barcoding as a complementary technique can substantially improve the level of resolution in gastric content analysis.

It has been 14 years since the rise of DNA barcoding technology. With the advancement and cost reduction of DNA sequencing technology and the gradual improvement in genetic databases, the emergence of next-generation sequencing (NGS) has further improved the resolution of analysis and given rise to methods for obtaining amplified subsequences of barcoded genes using high-throughput sequencing technology, i.e., DNA metabarcoding. DNA metabarcoding is based on the combination of DNA identification and high-throughput sequencing, where DNA fragments of an entire mixed sample are amplified, and then, high-throughput sequencing is used to automatically identify multiple species in the mixed sample in combination with biological information, which can reduce sampling effort and maximize the identification of semidigested/digested tissue residues. In terms of species identification, Harms-Tuohy et al. (2016) divided the stomach contents of *Pterois volitans* into two samples, digested (digest) and undigested, and sequenced their prey species separately using DNA metabarcoding. The resulting sequences were compared to GenBank and Barcode of Life Database databases, and of the 39 prey organisms identified by the undigested component, only four species could not be identified from the digest. Thus, DNA metabarcoding is a method that can conduct the "identification of highly digested stomach contents."

6.5.4 Main Factors Affecting Fish Feeding

6.5.4.1 Morphological Characteristics of Feeding Organs in Relation to Fish Feeding

The feeding habits of fish and their feeding behaviors are influenced by the morphological characteristics of their feeding organs and environmental and ecological factors such as food security and water temperature. Physiological activities such as spawning and overwintering, as well as the morphological characteristics of the feeding organs of fish, are closely related to their feeding patterns; however, studies on these topics are rare, and systematic theories and research methods have not been developed. For example, Chen Dagang et al. (Chen et al. 1981) used biomathematical methods to study the relationship between the morphological characteristics of the digestive organs of flounder and their feeding habits. The specific method involved selecting the typical quantitative indicators of fish digestive organs, such as the muzzle, head, mouth, intestine, and pyloric pendulum (mean values), and trait indicators, such as teeth, gill rakers, stomach, and anus (the distance between numbers was used to express the differences between various fish species), from which the information matrix x of fish morphological indicators was obtained, and then, the Euclidean distance (specimen point distance) between each fish species was calculated d_{ij}:

$$d_{ij} = \left\{ \sum \left[\left(x_{ij1} - x_{ij2} \right) / s_i \right] \right\}^{1/2}$$

where s_i is the standard deviation of the rows. The ecological types of the feeding fish were classified by cluster analysis of the distance from specimen sites.

6.5.4.2 Food Security in Relation to Fish Feeding

Food security, i.e., the availability of prey organisms in the environment, including the availability of prey biomass and the ability of consumers to capture and use it, is one of the main ecological factors affecting fish feeding. Fish and their prey live in a constantly changing environment. Therefore, natural selection of prey by fish can only be evaluated objectively by simultaneous sampling consumers and prey and comparing the composition of consumer stomach contents based on the composition of prey types in their environment. However, many of the tests needed for such studies are incomplete or immature, and the difficulties in sampling make it difficult to obtain sufficiently precise quantitative information so that the selection of prey by fish is often studied by the food selection index I in experimental ecology:

$$I_i = (r_i - p_i)/(r_i + p_i)$$

where I_i is the selection index of fish for prey i, r_i is the proportion of prey organisms in the stomach contents of fish, and p_i is the proportion of the same prey organism as that in the environment. I_i values range from $-1 < I_i < 1$. For $I_i > 0$, fish actively select prey i; for $I_i < 0$, fish avoid prey i.

In addition, some scholars use another selectivity index, I:

$$I = \frac{r_i - p_i}{p_i}$$

where I is the selection index, r_i is the percentage of a component in the food, and p_i is the percentage of the same component in the food base. Size is used to determine how selective a particular fish is for a particular prey. When the selection index is 0, it indicates no selectivity for this component; a positive value of the selection index indicates selectivity; and a negative value of the selection index indicates dislike for this food.

6.6 Fullness and Fat Content

6.6.1 Fish Fullness

Fish fullness is a measure of weight gain or loss in fish, and it is an indicator of how well the fish feed at different times and in different waters. The formula for calculating fullness is the following:

$$Q = \frac{W \times 100}{L^3}$$

where Q is fullness, W is body weight in grams, and L is body length in centimeters.

This is one of the indicators of fish growth expressed in terms of fish weight in relation to cubed body length. This indicator assumes that fish do not change body size as they grow. A change in the fullness factor indicates a change in the relationship between fish length and weight. An increase in body weight with a constant body length increases fertility; conversely, a decrease in weight indicates a decrease in fertility.

The fertility factor is actually the ratio of two measures, i.e., the volume of the fish (proportional to the weight of the fish) to the cubic product of the length of the fish. Therefore, when comparing the fertility of fish at different times and in different waters, separate calculations should be conducted for each age group and each length group, and the values should be compared for the same age and length groups.

In addition, the maturity of the fish gonads and the fullness of the gut, among other things, affect fatness and cause errors and variations. To eliminate this effect, net body weight is used as a proxy for total fish weight. However, after the gut is removed, some of the body fat will also be removed, which affects the correctness of the fullness determination. To address this problem, it is best to calculate both fullness levels simultaneously for correction.

6.6.2 Fat Content

Lipid content is the amount of fat stored in a fish and an indicator of how well a fish is feeding on nutrients at different times and in different waters, and it is more accurate than fullness.

Fat in fish is a nutrient that gradually accumulates in the body after the assimilation and digestion of food. The accumulation of fat in fish varies with the development of the individual and different life stages. Immature juvenile fish grow rapidly, food taken from the outside world is mainly used for development after assimilation, and very little fat accumulates in the growing body. With the gradual growth of the fish body, body fat gradually accumulates. Before and after sexual maturity of fish, the body fat content is high and often changes with the development of the gonads; generally, when spawning is over and after the resumption of feeding, the gonads and the amount of fat grow at the same time. However, when feeding stops, the amount of fat gradually decreases as the gonads continue to grow, so near the time of spawning, the lipid content decreases due to a shortage of nutrient sources as a result of reduced feeding; the nutrients accumulated in the body are converted to gonad development. The fat content of fish is also related to seasonal changes. Generally, in the late stage of feeding, body fat content increases, and during the overwintering period, because of stopped or reduced feeding, body fat is constantly converted into energy and used for gonadal development; thus, fat content gradually decreases.

The lipid contents of similar species and the same species are related to the characteristics of their habits, in addition to differences in physiological conditions and life stages, with groups that have long migratory routes having higher lipid contents. In the case of overwintering groups, metabolic intensity is reduced because they stop feeding for a certain season, so their body lipid content remains high.

Fish lipid content is usually determined by visual inspection (lipid content classes), chemical determination, and specific gravity estimation. Specific methods of determination can be found in relevant reference books such as the *Aquatic Resources Survey Manual* (1981).

References

Aguilar R, Ogburn MB, Driskell AC, Weigt LA, Groves MC, Hines AH (2017) Gutsy genetics: identification of digested piscine prey items in the stomach contents of sympatric native and introduced warmwater catfishes via DNA barcoding. Environ Biol Fish 100:325–336

Amundsen PA, Gabler HM, Staldvik FJ (1996) A new approach to graphical analysis of feeding strategy from stomach contents data: Modification of the Costello (1990) method. J Fish Biol 48:607–614

Chen XJ (2014) Fisheries resources and fisheries oceanography. Ocean Press. (In Chinese)

Chen DG, Liu CA, Zhang SB (1981) A comparative analysis of the digestive organs and feeding characteristics of the flounders in the Yellow and Bohai Seas. J Shandong Ocean College 01:87–106. (In Chinese)

Chen XJ, Liu BL (2017) Biology of fishery resources. Science Press (In Chinese)

Dale JJ, Wallsgrove NJ, Popp BN, Holland KN (2011) Feeding ecology and nursery habitat use of a benthic stingray determined from stomach content, bulk and amino acid stable isotope analysis. Mar Ecol Prog Ser 433:221–236

Harms-Tuohy DA, Schizas NV, Appeldoorn RS (2016) Use of DNA metabarcoding for stomach content analysis in the invasive lionfish (*Pterois volitans*) in Puerto Rico. Mar Ecol Prog Ser 558:181–191

Lindeman RL (1942) The trophic-dynamic aspect of ecology. Ecology 23:399–418

McClelland JW, Montoya JP (2002) trophic relationships and the Nitrogen isotopic composition of amino acids in plankton. Ecology 83:2173–2180

Pethybridge H, Daley R, Virtue P, Butler ECV, Cossa D, Nichols PD (2010) Lipid and mercury profiles of 61 mid-trophic species collected off south-eastern Australia. Mar Freshw Res 61(10):1092–1108

Sakaguchi SO, Shimamura S, Shimizu Y, Ogawa G, Yamada Y, Shimizu K, Kasai H, Kitazato H, Fujwara Y, Fujikura K, Takishita K (2017) Comparison of morphological and DNA-based techniques for stomach content analyses in juvenile chum salmon Oncorhynchus keta: a case study on diet richness of juvenile fishes. Fish Sci 83:47–56

Yang RB, Xie CX (2000) A review of the contents and methods of fish feeding ecology. Water Res Fish 03:1–3. (In Chinese)

Check for updates

Fish Schooling and Migration

7

Xinjun Chen and Bilin Liu

Abstract

Fish schooling and migration is one of the important living habits of fish and one of the basic contents of fishery biology research. The fish living in the sea generally have the habits of schooling and migration, which is the result of their adaptation to the environment (including the living and nonliving environment) in the long-term life. The main issues of concern to fisheries scientists and producers are when, where, and in what areas fish schooling occurs, how long fish schooling lasts and the size of the stock, etc., what are the marine environmental conditions for fish schooling, when the fish begin to migrate, and how the migration routes are distributed. Therefore, the purpose of studying fish schooling and migration is to master the law of fish schooling and migration, as well as the response to the marine environment, so as to provide a basis for the rational development and utilization of marine fishery resources. In the marine fisheries, most fish schooling are exploited, so it is of great practical significance to study the behavior of fish schooling and migration. In addition, through the study of fish schooling behavior, the methods of artificial schooling or controlling fish schooling behavior, such as tuna purse seine drifting, can be found to improve fishing efficiency and economic benefit. This chapter describes in detail the concept of fish schooling and the cause of formation and the type of fish schooling, discusses the structure of fish schooling and the law of its change, and introduces the concept, the type, the mechanism, and the biological significance of fish migration. At the same time, combining the latest research results at home and abroad, this chapter introduces the research methods and cases of fish migration, which provides a basis for understanding and mastering the distribution of fish schooling and migration.

Keywords

Fish schooling · Fish migration · Tagging

Abbreviations

EEZ	Exclusive economic zone
GIS	Geographic Information Science
TN	The resource of marked released fish
PTEF	Polytetrafluoroethylene
SST	Sea surface temperature
X	The number of fish marked in a marked release
Y	The number of fish recaptured with markers
Z	The total number of fish caught during the fishery information period

X. Chen (✉) · B. Liu
College of Marine Sciences, Shanghai Ocean University, Lingang Newcity, Shanghai, China
e-mail: xjchen@shou.edu.cn; bl-liu@shou.edu.cn

X. Chen, B. Liu (eds.), *Biology of Fishery Resources*, https://doi.org/10.1007/978-981-16-6948-4_7

7.1 Schools of Fish

7.1.1 Fish Schooling and School Types

Schooling is a phenomenon where a number of individual fish in the same physiological condition and with common life needs assemble into groups to live together. Under different life stages and different marine environmental conditions, the sizes and forms of fish schools are different. Generally, fish schools can be divided into four types, feeding, reproductive, overwintering, and temporary schools, depending on the reasons for their creation.

(1) Feeding schools. A school of fish that predate their favorite prey for feeding purposes, according to their feeding habits, is called a feeding school. The fish in a feeding school have the same feeding habits. Generally, fish of the same species with the same feeding habits are generally similar in length; fish of different species tend to gather together for the purpose of feeding on the prey in the system as well. The density of a school of feeding fish depends largely on the extent and amount of prey and environmental conditions.

As fish fertility increases and environmental conditions change, fish reorganize; fish distributed in the tropics and subtropics form reproductive schools when they reach gonadal maturity or during repeat maturity stages, while fish distributed in the temperate and boreal zones form overwintering schools due to changes in environmental temperatures.

(2) Reproductive schools. Any group of fish formed by the confluence of individuals with mature gonads is called a reproductive or spawning school. The fish in this type of school generally have the same body lengths and are at the same stage of gonadal development, but they are not necessarily the same age. In addition, in comparison to other schools, reproductive schools are denser, more concentrated, and more stable.

(3) Overwintering schools. A group of fish that gather together to find a new environment with suitable habitats based on changes in environmental temperature conditions is called an overwintering school. Any fish of the same species with similar fertility, not necessarily individuals of the same age and length, may school for overwintering. During migration to overwintering grounds, many fish schools are divided into smaller schools according to their fertility, but when they reach the overwintering grounds, most of the smaller schools gather to form larger schools, which are very large in number. The fish at the overwintering grounds school, and depending on their feeding habits and fertility, they may stop feeding or reduce their feeding. Overwintering schools are generally dense, but the density of fish varies depending on environmental conditions.

(4) Temporary schools. A temporary concentration of schools caused by sudden changes in environmental conditions or by encounters with aggressive fish is called a temporary school. In general, scattered schools of food-seeking fish or fish on the move, regardless of their life stage, often cause temporary concentrations of fish when they encounter sudden changes in environmental conditions, especially rapid changes in temperature and salinity gradients or when they encounter a large number of organisms that are distasteful and cannot be swallowed by fish, resulting in the presence of aggressive fish; these schools are temporary schools. When environmental conditions return to normal, they may again separate and live normally.

7.1.2 General Patterns of Fish Schooling

In general, the patterns of fish schooling are as follows: during the juvenile period, most individuals of the same species born at the same time in the same marine area assemble into groups, each individual in the group has exactly the same biological status, and the rhythms of subsequent biological processes are the same; this pattern occurs for most populations of fish. The growth rates of juvenile fish are not exactly the same among individuals, so some that have sufficient food, good nutrient absorption, fast growth, and early gonadal maturity will often leave the original group and join a group that was born earlier and has matured gonads; in the general population, the individuals with slower growth and later sexual maturity converge with the later-born group with similar gonadal maturity status; in addition, in the general population, most individuals with average growth and similar gonadal maturity status remain in the same basic group. The assemblage of fish that has been reassembled by the divergence of the basic population is called a shoal of fish (stock of fish). In this stock, the fish individuals are not necessarily the same age but have similar biological status, act uniformly, and are together for long periods of time. Fish in the same school may sometimes disperse temporarily into smaller groups in pursuit of food or to escape enemies, these smaller groups are temporarily separate, and they will reunite automatically once environmental conditions are suitable.

7.1.3 Role of Fish Schools

Although different scholars have different views on the role of fish schools and the role and biological significance of fish schools are not sufficiently understood and studied, they are considered to be important in terms of at least the following aspects.

(1) Enhance Fish Defenses

One of the most compelling hypotheses regarding the role and biological significance of fish schooling behavior is the defensive role of baitfish schools against predation. It is now generally accepted that schooling behavior not only reduces the probability of baitfish being detected by predatory fish but also reduces the probability that identified baitfish will be successfully killed by predatory fish. A school of thousands or even millions of fish may seem conspicuous, but in reality, a school of fish is no more likely to be detected by a predator fish in the ocean than an individual. Because of the absorption and scattering of light by suspended particles in the water, the visible distance of an object in the water is very limited; even in particularly clear water, the maximum visible distance of an object is only approximately 200 m, and this distance is independent of the size of the object. In fact, the maximum visible distance is much smaller. Fish have evolved over a long period of time as a form of society to develop schools that reduce the probability of baitfish being found, and other forms of defense have developed to reduce the probability of discovered baitfish being successfully consumed by predatory fish. Experiments have been conducted in aquariums and have shown that green cod juveniles acting alone are eaten by cod in an average of 26 s, while a school of green cod juveniles take an average of 2 min and 15 s to be eaten by cod.

In addition, schooling behavior helps fish escape from moving gear. When only part of a school of fish is enclosed by gear, often all of them can escape. Fishermen experiences show that a good catch is only possible if the entire school is enclosed. This is because the individuals in the school are very sensitive and react extremely quickly, and as soon as one fish changes direction due to fright, the whole school produces a coordinated turning movement almost simultaneously. Thus, the schooling of fish reduces danger and allows early detection of predators.

(2) Improve the Success Rate of Catching Prey and Feeding

Food relationships are the basic form of inter- and intraspecies biological contact between organisms. The schooling of fish makes it easier for them to find and locate food. It has been found that baitfish assemble in schools, but certain predatory fish also school. From this information, it can be concluded that schooling behavior also plays a role in the identification and predation of fish. However, to date, research on this issue remains scarce.

It is thought that when predatory fish form a school, the total number of sensory organs increases as does the search area. When one member of a school finds food, the other members can also hunt for it. The search area is greatest if the members of a school remain just within sight of each other. Thus, fish can find food more often and faster when in a school than when alone.

(3) Contribute to the Reproductive Success of Fish

Individuals whose gonads have matured gather together for the purpose of spawning to form reproductive fish schools to improve reproductive performance and reproduce offspring. Because reproductive fish groups have particularly high water temperature requirements and are often restricted to a certain range of water temperatures, schools are dense and effective in improving reproduction. For most fish, schooling becomes necessary for spawning, and the fact that many individuals gather together for spawning and mating plays some role in the spread of genetic factors. There is no doubt that schooling is decisively important for the reproduction of fish and the maintenance of populations.

(4) Facilitate the Enhancement of Fish Dynamics, etc.

Numerous studies have shown that in addition to their defensive, predatory, and reproductive roles and biological significance, fish schools have a variety of other roles in the lives of fish. From a hydrodynamic point of view, swimming in a group in water can decrease the energy consumed by each individual, and the vortex energy generated by the fish that are swimming can be used by the other fish that are following them; thus, each individual in the group can reduce a certain amount of swimming effort and continue moving forward. Compared to individual fish, schools are more resistant to adverse environmental changes. Grouping behavior not only enhances the resistance of fish to toxicants but also reduces the oxygen consumption of fish.

Overall, it is a serious mistake to limit considering the significance of schooling behavior to only one biological factor; the survival effects of fish schools are not the result of one biologically significant factor but should be the result of a combination of many. For example, the group mutualism effect, the chaos effect, mimetic behavior (pretending to be large numbers and large animals, etc.), the energetic effect, and other effects are combined to make schooling behavior beneficial to the survival of fish.

7.1.4 Behavioral Mechanisms of Fish Schools and Their Structures

7.1.4.1 Behavioral Mechanisms of Fish Schools

It has been shown that fish information is transmitted primarily through sound, gestures, water flow, chemicals, light flashes, and electric fields and that vision, lateral line sensations, hearing, smell, and electrical sensations may all play important roles in the formation and maintenance of fish schools, but there is no additional consistent accounts of the behavioral mechanisms that govern fish schools.

1) Role of Vision in the Schooling Behavior of Fish

Vision is the most important sensory organ of fish schools and plays an important role in schooling. In addition to vision, hearing, lateral line sensations, and smell are also closely related to schooling behavior. It is believed that vision has two main roles in fish schooling behavior: one role is that individuals lure each other through vision, and the other is that the swimming direction of the group is synced through vision. Luring is the first stage of schooling, enabling individuals scattered in any direction to focus on one direction and occurring mainly when the group is in a stationary state. Syncing direction, on the other hand, plays a role in ensuring the individuals gathered in one place face the same direction, maintaining sufficient space around each individual to sync their behavior, and it functions mainly in the mobile state of the group.

Thus, vision plays an important role in schooling behavior by providing a mutual attraction between fish that allows individuals within a school to lure and approach each other. It has been further shown that the visual system is an important sensory organ for maintaining the distance and orientation of most neighboring fish.

2) Role of Lateral Line Sensation in the Schooling Behavior of Fish

Most fish have a lateral line system on both sides of the body. Lateral lines can play a role in fish school formation, but most researchers believe that vision is more important than lateral lines. Others have suggested that lateral line sensation plays a role as important as vision in fish schooling behavior. Recent studies have further indicated that the visual system is an important sensory organ used to maintain distance orientation from the nearest neighboring fish, while the lateral line appears to be the most important sensory organ used to determine the speed and direction of neighboring fish. There is good evidence that both sensory organs are utilized together while swimming.

3) Role of Olfaction in the Schooling Behavior of Fish

Olfaction also plays a role in the schooling behavior of fish. It was found that the skin exudate of live and dead *Misgurnus anguillicaudatus* gave the same lure effect to their companions, and the *M. anguillicaudatus* skin exudate did not cause the companions to react with fear. The sense of smell is important for fish schooling when vision in *Phoxinus* cannot play a schooling role.

The above analysis suggests that the schooling behavior of fish is often achieved by information coming from more than one sensory source. Thus, in addition to visual, lateral line sensation, and olfactory senses, additional sensory systems may be involved in schooling and may be better understood when the complex behavior of fish is fully known.

7.1.4.2 Types of Fish School Structures

The study of fish school structure and its types is important for further elucidation of fish schooling behavior and determining fish stocks and fishing forecasts. The study of fish school structure needs to be considered from two aspects: first, the external school structure, such as the shape and size of the school, and second, the internal school structure, such as the species composition of the school, body length composition, swimming style, spacing, and speed of each individual.

In terms of the external structure of fish schools, the shape, size, and color of fish schools are different for different species of fish. Even for the same species, the external structure of a school will change with time, location, physiological state of the fish, and environmental conditions, especially for pelagic fish. For example, *Scomber* schools distributed in the northern China Sea area can be grouped into nine shapes: triangular, monoclinic, crescentic, tricuspid, flush, duck egg, square, round, and dumbbell (Fig. 7.1).

7.2 Migration of Fish

7.2.1 Description and Types of Fish Migration

7.2.1.1 Description of Fish Migration and Its Role

Due to genetic factors, physiological habits, and environmental influences, fish undergo periodic, directional, and regular school movement, a phenomenon called migration. Migration is an adaptive property of fish to expand their ranges and habitats to ensure the survival of species and increase the number of species, and it generally occurs on an annual basis. Migration is at the same time a social behavior, from one

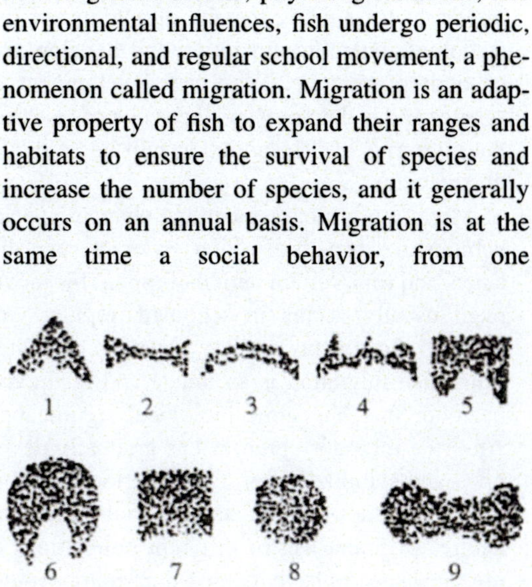

Fig. 7.1 Shapes of *Scomber* schools in the northern Yellow Sea (Chen 2014)

environment to another, a need, and an adaptation of species. Usually, fish migrations move along certain routes, and the paths through which they travel are called migration routes.

Understanding the migration patterns of fish is of great importance in fishery production. In certain seasons each year, some fish, especially some economically important fish, will migrate in schools, and their travel routes and the places where they gather to spawn, find prey, and overwinter often form concentrated fishing sites, forming fishing grounds. Therefore, it is important to study and determine the distribution of fish migrations and their patterns to forecast fishing seasons and fishing grounds, establish fish breeding reserves, improve the effectiveness of fisheries production and resource conservation and management, and release and increase fish stocks.

At present, it is an indisputable fact that human beings have seriously damaged the environment, a large number of aquatic environments have been polluted, many rivers have been blocked by hydropower construction, the survival of known migratory fish and unknown fish that may have migratory characteristics is seriously threatened, and some fish have even disappeared. Understanding fish migration mechanisms is important for guiding the appropriate use of fish resources, scientific utilization and development of hydropower, etc. With the development of science and technology and the scope of research, the mysteries of fish migration will be gradually uncovered.

Migration is a movement that changes species habitats in a certain direction, over a certain distance, and for a certain period of time. This movement usually occurs in schooled, regular, and periodic movements and has a genetic character. Thus, fish migration is an innate and instinctive behavior that has some biological significance. After the migratory process has gradually developed and stabilized over a long period of evolution, it becomes consolidated due to its hereditary nature. Different fish or different populations of the same species have their own migratory routes and certain reproductive, feeding, and overwintering sites due to different migratory heritable traits, which are the result of natural selection and are quite stable and cannot be easily changed.

Generally, fish can be divided into two main groups: migratory fish and resident/straggle fish. For most fishes, migration is an integral part of their life cycle. Only a relatively small number of fish, such as species of Gobiidae, regularly settle in one place and do not move over longer distances. Some species, such as some salmonids, migrate only as adults that have reached sexual maturity, and juveniles swim from the spawning grounds to the feeding grounds and then live there until sexual maturity without moving longer distances. In general, fish migration follows the life cycle of juveniles from spawning grounds → fattening grounds → feeding grounds → spawning grounds, while adults migrate directly from spawning grounds to feeding grounds and from feeding grounds to spawning grounds in such a cycle. For temperate fish, there is another overwintering migration due to the thermoregulatory effect of the fish, which involves a certain requirement for water temperature.

7.2.1.2 Types of Fish Migration

1) There are two main categories of migration, active and passive, depending on the dynamics of the migration. The migration of fish by their own motility is called active migration. For example, active migration includes migration to spawning grounds when approaching sexual maturity, migration to overwintering grounds when reaching a certain level of fertility, migration to feeding grounds after reproduction or overwintering, etc. The floating eggs, larva, or juveniles of fish are often carried by currents to faraway places because of their weak motility, and this movement is called passive migration.
2) Depending on the direction of migration, it can be divided into horizontal and vertical migration. Horizontal migration can be divided into landward migration, which refers to the movement of fish from an ocean into rivers, from an ocean to a coast, or from the lower reaches of rivers to their upper reaches, and off-land migration, which refers to the movement of

fish down a river, from a river to an ocean, or from a coast to an ocean. In addition, it has been noted in recent years that many fish species migrate from some of large inland lakes to their incoming rivers for reproduction; before winter, they migrate from the shallow waters of these rivers to the deeper waters (deep pools or caves) and even to the groundwater of caves connected to the rivers for overwintering. These types of migration also fall under the category of horizontal migration. Vertical migration refers to the vertical movement of fish between the upper and lower layers of a water column; some flounders and gobies switch from planktonic to benthic prey at a certain stage of their development, and such migration should also be classified as vertical migration.

3) Depending on the nature of a migration, it can be divided into reproductive, feeding, and overwintering migrations.

A) Reproductive migration (spawning migration; breeding migration). Reproductive migration is the movement from a feeding or overwintering ground to a spawning ground. Reproductive migration occurs when fish gonads mature because the gonads secrete sex hormones into the bloodstream, stimulating the nervous system and leading to the requirement of fish spawning and reproduction, and fish often gather in groups for fish spawning, offspring growth, development, and migration. Spawning migrations of fish are usually classified into three types, namely, anadromous, downstream, and landward migrations, depending on the migratory path and the ecological environment in which they spawn.

Landward migration is the migration from ocean depths to shallow coastal waters, and most fish are carrying out this type of migration. Anadromous migration refers to growing in the ocean and swimming upstream to spawn when mature, and fish that carry out this type of migration include Salmonidae *Tenualosa reevesii* and *Oncorhynchus keta* that swim from oceans into rivers and swim upstream to their spawning grounds to reproduce. Downriver migration refers to growing in rivers and swimming to oceans to spawn when mature; for example, *Anguilla japonica* enter oceans from rivers to their spawning grounds to reproduce, and their migration direction is opposite to that of cannabis salmon.

Reproductive migrations are characterized by (i) fast swimming speeds, long distances, and less influence from the environment. If the speed and direction of reproductive migratory fish are known in advance, then the next fishery and fishing period can be predicted based on the current fishing conditions. (ii) During reproductive migrations, splitting is most evident, usually in sequential groups by age or body length. (iii) During reproductive migrations, the gonads undergo dramatic changes, and the differences before and after are obvious, in terms of development, volume, and weight. (iv) The destinations of reproductive migrations are spawning grounds, which are in a certain ocean area each year but are subject to some changes under the influence of hydrographic conditions (e.g., changes in temperature and salinity).

B) Feeding migration. Feeding migration is the movement from spawning or overwintering grounds to feeding grounds. After overwintering, sexually immature fish and adult fish that have consumed substantial energy through reproductive migration and reproductive activities swim to the prey-rich ocean areas to intensively seek prey, grow and fatten, recover strength, accumulate nutrients, and prepare for overwintering and reproduction in the next year.

The purpose of the migration is to find prey, so the route, direction, and period of this migration are more variable, far from having a more stable range such as with reproductive migration. The main factors that determine feeding migration characteristics are nutritional conditions, while hydrological conditions (temperature, salinity, etc.) are secondary factors. Changes in the distribution and movement of prey govern the

Fig. 7.2 Migration relationship diagram for various migratory species (Chen and Liu 2017)

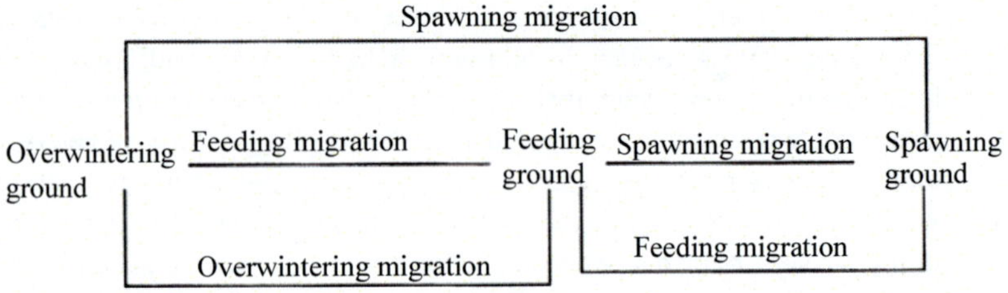

dynamics of the prey. After the fish have consumed many prey organisms, if the density of prey decreases to a certain level, and when the energy consumed by feeding on prey exceeds the accumulation of energy, then the fish will continue to migrate in search of new prey. The migration time and space tend to vary with the distribution of the number of prey organisms. Therefore, the distribution and movement of prey can generally correctly determine the changes in fishing grounds and fishing periods. For example, on the northern coast of China, *Trichiurus lepturus* likes to eat *Ammodytes personatus*, *Setipinna tenuifilis*, etc. Every year, after these baitfish arrive at the striped bass fishery, after approximately 10 days, a large number of striped bass can be caught. In another example, many Chinese fish that spawn in spring and summer generally seek prey in nearby sea areas after spawning.

C) Overwintering migration. Overwintering migration refers to the movement from feeding grounds to overwintering grounds. Fish are variable temperature animals and are very sensitive to changes in water temperature. Fish have different temperature ranges, and when the ambient temperature changes, fish will move in schools in pursuit of waters suitable for their survival; this movement is called overwintering migration.

The characteristics of overwintering migration are as follows: (1) Fish usually move in the direction of gradually increasing water temperature during their overwintering migration. Therefore, the migration direction of Chinese marine fishes generally proceeds from north to south and from shallow waters to deep waters. (2) During overwintering migration, fish usually reduce or stop feeding, relying mainly on the nutrients accumulated in their bodies during the feeding period to supply the energy they need. Therefore, the distribution and changes in prey during this period generally do not govern the actions of fish. (3) Only when fish reach a certain level of abundance and lipid content is it possible for them to carry out overwintering migrations, so the biological state of a fish is the basis for migration. Fish are prompted to being their overwintering migrations when they reach a certain biological state and are stimulated by environmental conditions (e.g., a drop in water temperature), so changes in environmental conditions are the conditions for migration. Fish that have not reached a certain level of abundance and lipid content continue to find prey for fattening but do not engage in overwintering migration. Overwintering migration involves searching for warmer waters, so it is strongly influenced by water temperature conditions, especially the distribution of isotherms.

Reproductive, feeding, and overwintering migrations are interrelated, with the first part of a life cycle preparing a fish for the later part (Fig. 7.2). The transition to a migratory state is linked to certain biological states of fish, such as abundance, lipid content, gonad development, and blood osmolarity. Thus, the onset of migration in fish depends primarily on the biological state of the fish but also on changes in environmental conditions.

However, not all migratory fish species undertake all three migration types; some fish species carry out only reproductive and feeding migrations but do not carry out overwintering migrations. There are also fish species where the three migrations cannot be separated and crossover to varying degrees. For example, in the case of subspawning fish, small-scale feeding may already take place within spawning grounds; in the case of feeding migrations, they may be interconnected with overwintering migrations due to changes in prey biomass or seasons.

7.2.2 Mechanisms and Biological Significance of Fish Migration

7.2.2.1 Factors Affecting the Migration Process of Fish

The factors affecting the migration process of fish are very complex and include both internal and external factors, and fish migration processes are the result of the combined action of internal and external factors. That is, when the state of physiological activity of a fish reaches a certain level, there is a corresponding stimulation of environmental factors that contributes to migration.

1) Internal Factors

The dominant factors affecting the migratory processes of fish are internal factors, that is, changes in their biological states, such as gonadal development, hormonal action, and changes in fatness, lipid content, and blood chemistry. When the gonads develop to a certain extent, the secretory action of sex hormones causes the corresponding activity of the nervous system, which leads to the reproductive migration of the fish. A certain level of fertility and lipid content must be reached to initiate overwintering migration. Feeding migration occurs due to the need for food during reproduction or after overwintering.

Changes in the chemical composition of the blood and the mechanisms of osmotic pressure regulation in fish are also internal factors that affect the migration process. Before entering oceans, *A. japonica* have a progressively higher blood carbon dioxide content and thus an increased blood osmotic pressure, which makes entry into an ocean a physiological imperative. When salmonids enter freshwater, their blood osmotic pressure gradually decreases, the digestive tract atrophies, and feeding stops, which enables their reproductive migration to proceed actively.

Reproductive migrations will not begin in fish with poorly developed gonads, even if they have reached reproductive age. Similarly, the overwintering migration of fish will not begin even if winter has arrived if they have not yet reached a certain level of fertility and lipid content. Therefore, internal factors are the dominant factors affecting the process of fish migration, while changes in external environmental conditions play a stimulating or inducing role in migration.

2) External Factors

The fact that a fish has completed its migration preparation does not mean that its migration will start immediately. Usually, a fish that is ready for migration will start its migration only when it is stimulated by certain external factors, and at the same time, a migration that has already started is still subject to the influence of external factors. In addition, fish will often temporarily stop their migration activities or deviate from the annual migration route due to the presence of unfavorable external factors. Thus, external factors not only act as a signal or stimulus to cause the start of migration but also affect the entire process of migration.

There are many external factors that affect fish migration; however, the magnitude of their effects varies greatly, and they must be distinguished as primary and secondary factors. It is worth noting that this division between primary and secondary factors is not fixed but varies by species, and even if the same species is at different developmental and life stages, the primary and secondary external factors will be transformed into each other. In the period of reproductive migration, fish migrate mainly to seek suitable sites for reproduction, and in the process of swimming to spawning grounds, external factors such as water temperature, salinity, transparency, and flow rate often have had a more significant influence on their behavior; in the period of feeding migration, migration is mainly to find prey, so the main external factors determining the behavior of fish have been transformed into prey; in the period of overwintering migration, migration is mainly to

Table 7.1 Internal and external factors affecting the various types of migration (Chen 2014)

Type of migration	Key internal factors	Key external factors
Reproductive migration	Physiological condition has reached a certain point where the gonads have begun to mature. For example, salmonids enter rivers due to gonadal stimulation, changing body shape, body color, etc.	Stimulation by external environmental conditions, mainly temperature, which does not reach the required temperature, even if the gonads are fully mature and cannot ovulate
Overwintering migration	Fatness and fat content meet certain requirements	Water temperature decreases
Feeding migration	Starvation due to physical exertion from spawning or after overwintering	Distribution of prey organisms

find suitable overwintering sites, and fish gradually swim to ocean areas with higher water temperatures or deeper water with slower currents; thus, the external factors determining this type of migration are mainly water temperature and topography.

Taking the Chinese East China Sea *Trichiurus lepturus* as an example, the external factors affecting its overwintering migration in the Shengshan fishery are mainly water temperature, salinity, water mass, wind, current, transparency, and water color. The influence of water temperature is the most obvious; when the surface temperature of the Shengshan fishing ground drops to 20 °C, *Trichiurus lepturus* enter the fishing ground; when the water temperature is 13 °C, striped *Trichiurus lepturus* leave the fishing ground, the water temperature during the peak season is 15.3–18.6 °C, and the development of the fishing period can be predicted based on the forecasted water temperature. *Trichiurus lepturus* generally move south along the 30–40 m isobath, and the water color of the fishing ground is 11–14, which fishermen often call "white rice" water color. In the mixed area where there is a low temperature and low salinity level in the water system along the coast and a high temperature and high salinity content in the water outside the sea, the fish are dense. When the coastal current is strong, the migratory route is outward, and vice versa. Wind conditions also affect fishing conditions, with easterly winds strengthening the outer high-temperature, high-salt water system and inward migration routes, and the opposite scenario occurs with westerly winds. When there is a prolonged lack of high winds, the water temperature is vertically stratified, and striped bass have a decreased ability to school, which is not conducive to production; after a storm, the flow is clearly separated, and the fish are dense, which is conducive to production. If continuous storms occur, then the fish accelerate southward, and the fishing season ends early, which was also unfavorable to production. From the above information, it can be seen that exploring the various external factors affecting the overwintering migration of striped bass and determining the main influencing factors can serve as a scientific basis for fish forecasting, and these factors play a certain guiding role in production.

Thus, the factors influencing each type of fish migration are briefly summarized in Table 7.1; i.e., internal factors are dominant, while external factors are conditional.

7.2.2.2 Orientation Mechanisms During Fish Migration

Almost all fish migrate in schools. Migratory schools are generally composed of fish of similar lengths and biological status. The fish in a migratory school do not have a fixed leader, and the first fish will fall behind after a while and be replaced by others. Migratory schools usually have a certain shape, ensuring the most favorable dynamics of the school. The migratory adaptation of fish is a function of not only more favorable hydrodynamic conditions for movement but also the ease of orientation during migration. The fact that the size of migratory schools varies from species to species is undoubtedly related to ensuring the most favorable migration conditions.

Fish are able to use their sensory organs to orient themselves and successfully complete migrations that are sometimes thousands of kilometers long. There is not yet a satisfactory answer to why fish migrate in a certain direction and on a certain route and spawn in the same place. The following main factors are generally considered in response to those questions:

1) Water chemistry factors. Water chemistry, especially salinity, is an important factor affecting fish migration because changes in salinity in the water cause changes in osmotic pressure in fish, which leads to an excitation of the fish nervous system. In addition, changes in water quality play a large role in migration.

Numerous studies have shown that salmon rely on their sense of smell to successfully locate their original birth rivers for spawning, and the odor of the water in their birth rivers plays a guiding role. Some believe that gradient distributions of salinity and dissolved oxygen levels are also often used for orientation during fish migration and that fish may be able to sense whether they are leaving or approaching a coast based on the gradients of these chemical factors, thus allowing migration to occur successfully. Similarly, fish may use the distribution of water temperature gradients for orientation. Others, however, argue that gradients of salinity, dissolved oxygen content, and water temperature are too small to be sensed by fish sensory organs and may therefore be of little relevance to migratory orientation.

2) Currents. Fish sense currents through their sensory organs, mainly the eyes, and certain skin sensory organs, such as the lateral line, and the direction of fish movement can be indicated by current-sensing stimuli from the lateral line. In general, the long-distance migration of fish is indicated by current as an indicator of orientation, and the passive migration of juvenile fish depends entirely on current. It is also now generally accepted that fish are capable of relying on sensing the current for migration orientation, which occurs in anadromous fish as well as in marine and freshwater fish. For example, *Gadus morhua*, *Clupea harengus*, and *Oncorhynchus keta*, which swim into the Heilongjiang River to spawn, swim forward in accordance with Amur currents once they enter the Sea of Okhotsk. *Pseudaspius leptocephalus* and *Rutilus rutilus* also orient themselves according to currents.
3) Fish tropism. Under certain conditions, fish generally have positive tropism. Some studies have suggested that natural currents in the sea formed by the Earth's magnetic field may have a role in the migratory orientation of fish. However, it has been suggested that the strength of currents capable of inducing positive electrophilicity in fish is four to nine times greater than the strength of natural currents measured in the sea, so it seems unlikely that fish are oriented according to natural currents.
4) Temperature. Latitude plays a large role in the direction and route of fish migration. Fish are variable temperature animals and have particularly stringent temperature requirements when spawning and therefore follow certain isotherms.
5) Topography. Many species of fish may also rely on shoreline and ocean bottom shapes for orientation during migration, which may be related in part to water pressure perception.
6) Historical genetic factors. Fish migration is hereditary, and this heritability is specific to each species and each population so that different populations have different characteristics. Heritability is a characteristic that arises from ancestors of a species through a long historical process of continuous selection and exists within the nervous system, and the differences caused by evolutionary history are also involved in the process of heritability formation so that a specific behavior, which is instinctive, is produced when stimulated by internal and external conditions. Genetic factors are one of the main reasons why fish undertake annual spawning migrations, feeding migrations, and overwintering migrations.
7) Cosmic factors. Environmental and hydrographic factors play an important role in the

direction of migration, particularly cyclical changes in currents, which lead to periodic migrations of fish. The cyclic variation in sea currents is associated with cyclic variations in geophysical and cosmic aspects, especially with variations in the heat received from the Sun. Radiation from the Sun is related to the activity of sunspots, which have an 11-year cycle. When sunspot activity increases, thermal radiation increases, and oceans absorb enormous amounts of heat with resulting increases in water temperatures, thus affecting warm current temperatures and flow potentials during a year, which has a direct impact on the development and migration of marine fish.

7.2.2.3 Biological Significance of Fish Migrations

Since fish migration has gradually developed over a long evolutionary process and is the result of the long-term adaptation of fish to external environments, it will inevitably have a certain biological significance. It is now generally accepted that fish migration ensures favorable conditions for the survival and reproduction of populations. Reproductive migrations are adaptations that ensure that eggs and young eggs are provided the best possible developmental conditions, especially as an adaptation to defend against aggressive animals in early developmental stages. Feeding migration facilitates the availability of abundant prey organisms for fish, thus enabling rapid growth and development of individuals and the maintenance of large populations. Overwintering migrations are specific to species that require overwintering, ensuring that overwintering fish have the most favorable abiotic conditions and adequate defenses against predators at low levels of mobility and metabolic intensity. Overwintering is an adaptation that ensures the survival of the population during seasons that are not conducive to activity. Overwintering is characterized by reduced activity, complete cessation or greatly reduced intensity of feeding, and reduced metabolic intensity, relying mainly on accumulated energy in the body to maintain metabolism.

For example, the feeding and reproductive migrations of pelagic fish generally occur from the outer sea to the coastal zone. The coastal zone has warmer water, strong currents, and abundant nutrient organic matter, ensuring the availability of prey. Because the coastal zone is narrower, it is much better for males and females to meet during fish reproduction than the vast ocean. Because of the rapid increases in temperature and the availability of sufficient prey, the egg development period can be shortened, allowing for an earlier migration from a dangerous period and hatching of young individuals. On the other hand, currents flowing from the mainland to the ocean can have an effect on these fish. However, the coastal zone is not favorable to fish at all times. When it becomes cold, the water temperature can drop rapidly, and food can be reduced. These conditions result in overwintering migrations to certain ocean depths.

7.3 Methodologies for Studying Fish Migration

Studying the distribution of fish migrations is a major component of studying the biology of fishery resources. Its aim is to understand the migratory distribution of fish and the relationship with the marine environment.

The main methods for studying the distribution of fish migrations are exploratory surveys, mark-release methods, statistical analysis of catches, direct reconnaissance using instruments, use of fish biological indicators, and trace element analysis, all of which have advantages and disadvantages. If a large number of survey vessels can provide detailed, accurate, and continuous production records over a long period of time, then a statistical analysis of the catch method is the most practical method, but often due to the interests of various parties and the abilities of the crew and other factors, this desired purpose is often not achieved. If fishery resource survey vessels are sent out specifically to obtain data, then the data obtained are accurate and relevant but costly, limited in scope, and time-consuming to obtain. The mark-release method is a more traditional method, the results are the most intuitive and effective, and it is relatively inexpensive and simple. With the application of satellite remote sensing technology and the development of microelectronics, the mark-release method has been improved, especially with the emergence of data storage signs and detached satellite signs. This section analyzes the mark-release method, statistical analysis of catch method, and trace element analysis method.

7.3.1 Statistical Analysis of Catches

Fishing records from survey vessels are collected in large quantities over a long period of time; catch statistics are conducted by fishing area, species, and month; and the statistics are used to map the catch distribution of each fishing area separately by species. Based on time-series catch distribution maps, the migration routes and distribution ranges of fish can be analyzed and inferred. This work can be carried out continuously over a long period of time, and fishing charts of various economic fish species can be drawn, which provides an important reference for the analysis of fishing grounds and fishing periods. The advantages of this method are its low cost and effectiveness, but the disadvantages are that it requires a long series of fishing logs, particularly precise operating vessel positions, the yield of various species, and their biological characteristics, and in addition, it is difficult to analyze the relationship between fish migration and the environment independently.

With the development of Geographic Information System (GIS) and marine remote sensing technology, long-term historical production statistics, combined with remote sensing information, such as surface temperature, chlorophyll and sea surface height, and catch distribution maps at different times, can be drawn (Fig. 7.3) to determine fish migration distribution patterns as well as suitable environments.

7.3.2 Tagging

7.3.2.1 Concept of Mark-Release Method

The mark-release method involves attaching a tag or mark to a captured fish or installing an electronic marker on a captured fish, releasing it back into the sea, and then analyzing it based on release records and recapture records. Mark-release methods have been an important part of aquatic resource science for a long time. The number of targets that have been marked and released has been increasing, and the use of this method has been expanding. Currently, the mark-release method has been used for various aquatic animals, such as crabs, shrimp, shellfish, and cetaceans, and is carried out on economic fish as well. Mark-release methods can be divided into two main categories according to the methods used, namely, the marking method and the tagging method. The marking method was one of the earliest methods used and is a method that involves marking the original organs of a fish, such as by the total or partial removal of fins. The tagging method involves attaching special markers to an aquatic organism, and the markers usually indicate the marking unit, date, and location. Tagging is the most prominent method used in modern mark-release efforts and can be divided into in vitro marking, in vivo marking, biotelemetry marking, data storage marking, and detached satellite marking.

7.3.2.2 Significance of the Mark-Release Method

Analyzing aquatic organisms marked for release and recapture over a considerable period of time, based on the time and place of release and recapture, provides information on the movements and growth of fish in the water and is a common method for surveying fisheries and studying the distribution and growth of fish migrations. This information is recorded and can be used as a reference for estimating resource holdings. Mark-release methods are of great importance for fisheries production, mainly in terms of the following:

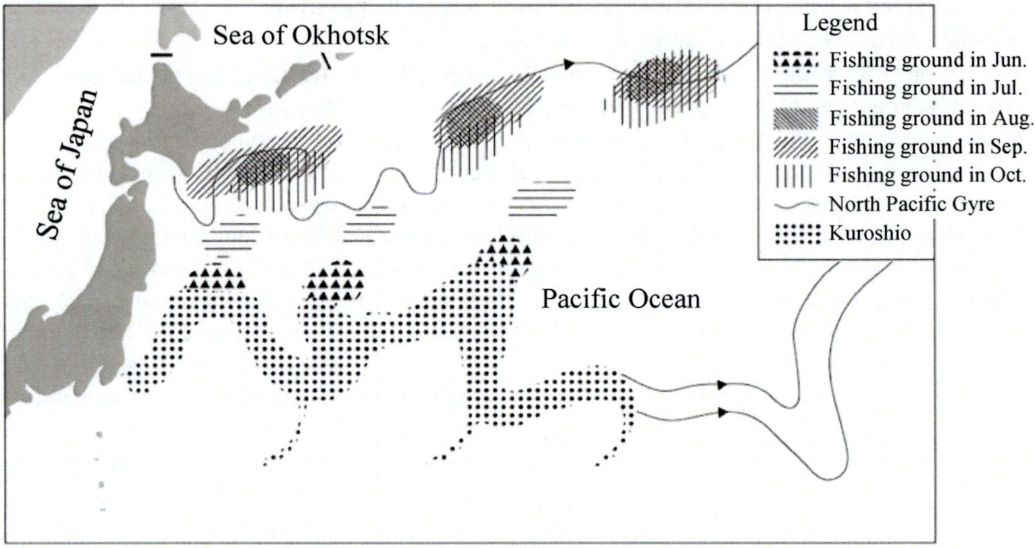

Fig. 7.3 Schematic diagram of migrations of neon flying squid based on catch and operational locations in the North Pacific Ocean (Chen and Liu, 2017)

1) Determination of the direction, route, speed, and extent of migratory movements of fish. A marked and released fish (or other aquatic organism), during its stock movement, is recaptured in a certain ocean area at a certain time so that the direction, route, range, and speed of its movement can be inferred in comparison with the time and place of the original release. This measure is the most effective way to determine fish migration directly. However, the extrapolation of the migration speed based on the distance from the release to the recapture site can only be used as a conceptual reference and cannot be used as an absolute migration speed.
2) Extrapolation of the growth rate of fish length and weight. The length and weight growth rates of fish can be extrapolated from records of length and weight measurements of marked fish at the time of release compared with those of recaptured fish over a significant period of time.
3) Extrapolation of approximate catch rates to estimate resource holdings. If a large number of fish are marked for release, then the number of fish returning to the original stock may be high, and the chance of recapture may be greater. If these fish are properly dispersed in the original stock, then the ratio of recaptures to releases will approximate the ratio of catch to resource. Thus, using the total number of fish marked for release and the total number of recaptures collected at a given fishing ground during a season as a basis and applying various corrections to the results of the releases, an approximation of the catch rate can be estimated, and in combination with the total number of fish caught, the resource holdings can be estimated, providing a reference for fishing and management purposes.

The calculation involves the following: let the number of fish marked in a marked release be X, the total number of fish caught during the fishery information period be Z, and the number of fish recaptured with markers be Y. The resource of marked released fish, TN, is

$$\mathrm{TN} = \mathrm{XZ}/\mathrm{Y}$$

4) he relationship between fish migration and the marine environment can be analyzed, indicators of fishery formation can be explored, etc.

7.3.2.3 Mark-Release Approaches

1) In vitro tagging method (external tags). This is a commonly used method of tagging releases. That is, a tag that has a clear color is attached or tied to an appropriate part of the exterior of the released fish. This method is a traditional and simple method with a number of drawbacks. However, it is relatively inexpensive to operate, and few valid data can be obtained.

When utilizing in vitro tags, issues such as the amount of resistance to fish movement in the water and corrosion of the material should be considered before marking the organism to be released may be achieved. Currently, small metal sign tags are generally used, with materials such as silver, aluminum, or plastic, followed by nickel and stainless steel, and sign and spike types are more commonly used today (Fig. 7.4). All sign tags should be engraved with the representative font size and tag number of the release agency, and the release location and time should be recorded in order on the record sheet of the sign release so that it can be used for basic data for marking the fish after recapture. Marking sites vary according to the size of the fish (Fig. 7.5).

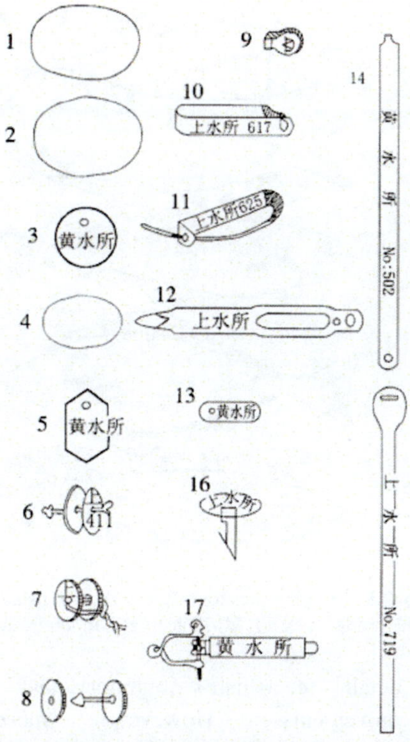

Fig. 7.4 Types of tags. 1–5. Tag types; 6–8. Buckle tag types; 9–12. Clip-on tag types; 13. Body markings; 14–15. Belt type; 16. Lift-off type; 17. Hydrostatic type

2) Isotope marker method. Radioactive isotopes with a long radiation period (usually 1–2 years) that are not harmful to fish are introduced into fish as markers, and the most commonly used isotope testers are used to detect the recaptured marked fish. The most commonly used isotopes are P^{32} and Ca^{43}. There are two methods of introducing radioisotopes into fish: feeding the fish with food containing the isotope or placing the fish in water dissolved with the isotopic material and infecting them directly. This method of release is simple to perform, but recovery is more difficult because the marked fish are difficult to detect.
3) Biotelemetry marker method. This method takes advantage of remote sensing devices and involves installing an ultrasonic or electric wave generator onto a fish body as a marker, and after the marked fish is released, a test boat equipped with a receiver can be used to follow and record and continuously observe to determine the migratory route, speed, depth change, diurnal activity pattern, etc. of the marker fish. This method is simple and allows a more detailed record of fish life history patterns but is generally not used for a long period of time.
4) Data storage marker. A data storage marker is a microcomputer-controlled recording device. This method involves placing a data storage marker inside the body cavity of a captured fish, and once the fish is released, the marker is activated every 128 s for a total of 675 times a day, recording water pressure, light intensity, and internal and external temperature data from four sensors. Each day, the marker uses the recorded quorum data to calculate the geographic location for that day with a fair degree of accuracy. Based on the information stored in the markers, researchers can determine the

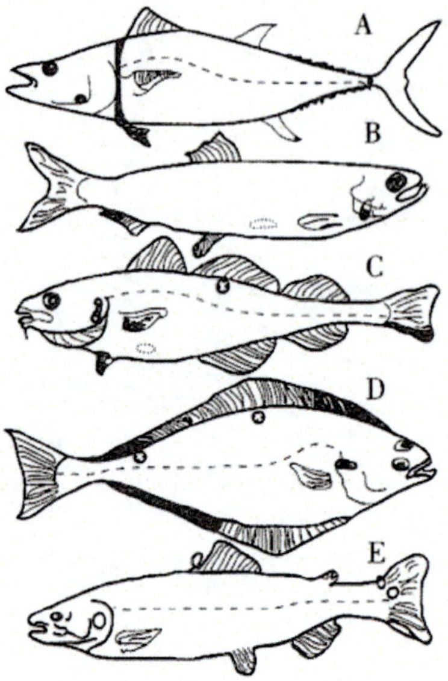

Fig. 7.5 Marking sites for different sizes of fish. (**a**) Tuna; (**b**) Herring; (**c**) Cod; (**d**) Plaice; (**e**) Salmon trout

details of a fish's migratory and vertical movements. However, successfully performing this step requires retrieving the marker when a fish is recaptured.

Installation of the data storage marker involves mounting a small fish marker in the body cavity, or a large fish marker is inserted in the dorsal muscle immediately adjacent to the first dorsal fin. Practical experience has also confirmed that the muscle insertion method is faster to complete and less dangerous than placement in the body cavity.

Data storage marker features are the following: body made of stainless steel, 52 g, 16 mm diameter, and 100 mm long and conduction rod made of polytetrafluoroethylene (PTFE), 2 mm diameter, 200 mm long, and a battery life of over 7 years.

7.3.2.4 Separate Satellite Sign

The main components of a detached satellite sign include a clock, sensors, a control storage device, an uplift control section, an energy supply device, and the housing. The functions of each device are briefly described as follows: the clock provides the time for the sign device; the sensors are used to obtain information on different environmental parameters, and the common configurations include temperature sensors, pressure sensors, and brightness sensors; the uplift control section can be used to control the release of the satellite sign from the fish body, which in turn also indicates the end of the release process, and it mainly consists of a floating ring and an antenna to ensure that the sign can communicate with the Argos satellite system. The energy supply unit is the power system of the marker, from the acquisition of stored data to communication with the satellite after surfacing, so this energy system has a high capacity and durable performance; the control storage unit is the central system of the marker, controlling the normal operation of the other parts above; the housing is usually composed of corrosion-resistant and high-pressure-resistant epoxy hydroxyl resin, with a streamlined shape to reduce the resistance to movement of the fish body. The main companies currently producing detached satellite signs are Microwave Telemetry and Wildlife Computers in the United States, both of which transmit data via Argos satellite system (Fig. 7.6).

Satellite marker release methods have been widely used with success to study the mass movements (migrations) of marine animals, such as marine mammals, seabirds, sea turtles, sharks, and tuna species, and the physical characteristics of their habitats (e.g., water temperature). The first satellite marker release of tuna species was conducted in North Atlantic waters in September–October 1997, when 20 bluefin tuna were released after being tagged with PTT-100 satellite marker tags that were set for release in March–July 1998. Seventeen of these were recovered, and the collected data were successfully obtained. Their recovery rate reached 85%. The average data recorded per sign was 61 days. Through this data, some valuable information was obtained, such as the vertical and horizontal distribution of tuna at different times of the year, migration direction and route, and habitat water temperature (Fig. 7.6). In addition, the data obtained from the satellite markers, combined with environmental data obtained from marine remote sensing, can provide the migration routes of marine animals (e.g., sea turtles) and their relationship with their habitats (Fig. 7.7).

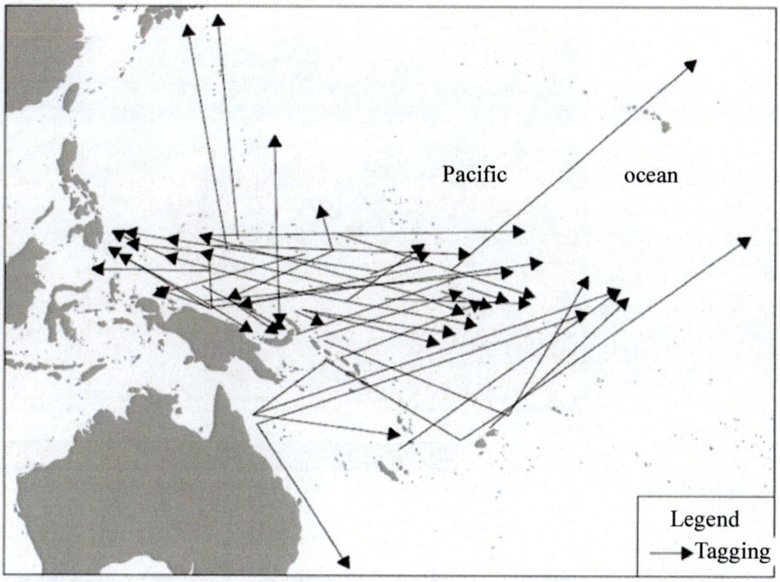

Fig. 7.6 Distribution of yellowfin tuna satellite markers conducted in the Central and Western Pacific Oceans (Chen and Liu 2017)

In addition, with the data obtained by satellite markers, such as movement location, date and time, and habitat environment, we can infer the migration distribution and movement speed of marked objects, diurnal vertical movement pattern, habitat pattern in different water layers and optimum water layer, the relationship between habitat distribution and temperature, as well as suitable water temperature and optimum water temperature, and we can also provide a scientific basis for more accurate assessments of fish resources and estimates of important habitats.

7.3.3 Trace Element Analysis Method

During the exchange of substances between fish and the external environment, chemical elements from the environment enter the body through respiration and ingestion and are then deposited in otoliths after a series of metabolic processes and circulation into endolymphatic crystals. These elements are deposited in the otolith in very small amounts after regressive transport through the body and are referred to as trace elements. Due to the noncellular and metabolically inert nature of otoliths, the chemical elements deposited in otoliths in the aquatic environment are essentially permanent as the fish and their otoliths grow in tandem. Fish otoliths record the characteristics of the water environment in which fish live throughout their life cycle, and changes in the water environment result in changes in otolith trace elements. The analysis of information related to the surrounding water environment and trace elements in otoliths can play an important role in life history analyses of fish migration, reproduction, and spawning, as well as in the reconstruction of habitats including elements such as temperature, salinity, and food.

(1) Life History Analysis

Analysis of trace elements in fish otoliths and their surrounding water environment and ecology allows the reconstruction of the life history of squid species such as *Todarodes pacificus*, *Gonatus fabricii*, and *I. argentinus*.

(2) Relationship Between Trace Elements and the Environment

The Sr content (equivalent to Sr/Ca) in fish otoliths is usually negatively correlated with water temperature. Although Sr/Ca in cephalopod statoliths is also negatively correlated with water temperature, the magnitude of the daily variation in water temperature they experience will affect the relationship between Sr/Ca and water temperature. Ommastrephids are mostly pelagic species with

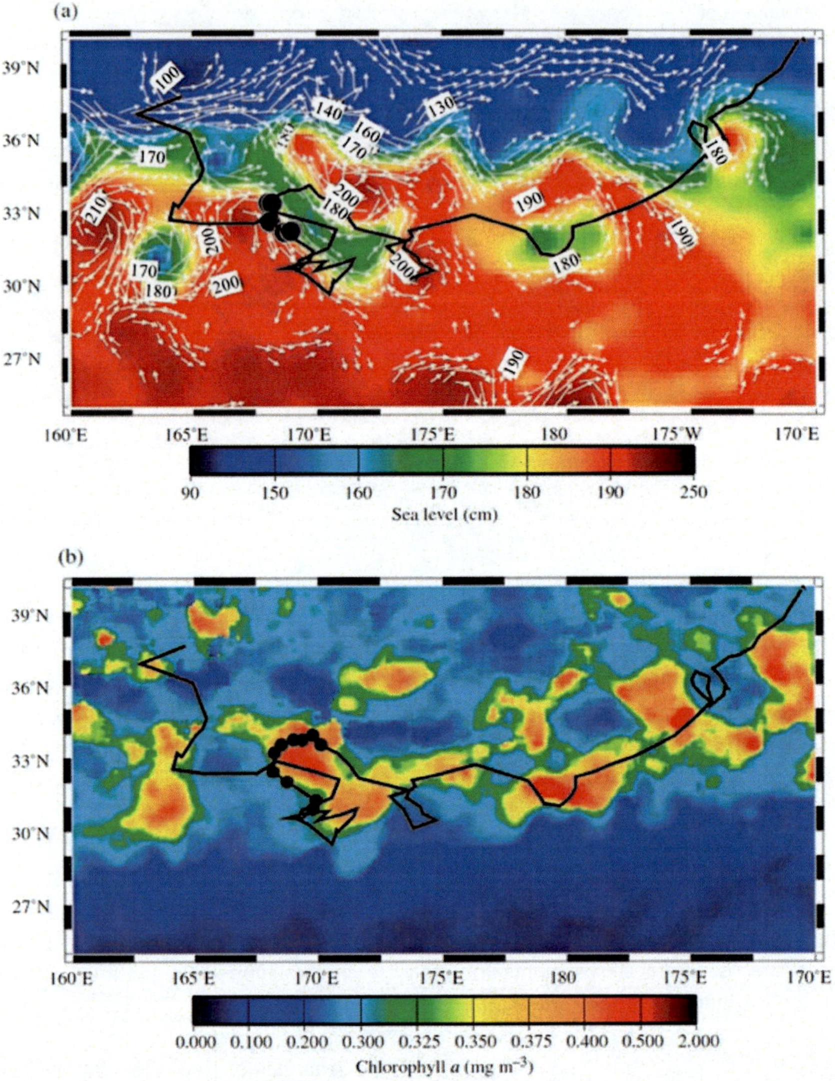

Fig. 7.7 Turtle migratory routes in relation to sea surface height (**a**) and chlorophyll (**b**) in the North Pacific Ocean (Polovina et al. 2004)

significant diurnal vertical movements (inhabiting deep cold water during the day and moving in warm surface water at night), and the large diurnal temperature differences result in nonsignificant Sr/Ca changes with overall water temperature. For example, a comparative study of *O. bartramii* statolith from El Niño and non-El Niño years in Peruvian waters found no significant difference in Sr/Ca despite a significant difference in water temperature in the 2 years. The statoliths of *Sepia officinalis* showed a negative correlation between Ba/Ca and temperature and a positive correlation between I/Ca and temperature, with the relationships $Ba/Ca = 26.89 - 0.574$ and $I/Ca = 1.49 + 0.244T$, respectively; however, Sr content was not significantly correlated with temperature, which is inconsistent with the results of other cephalopod studies and may be due to differences between field and indoor environments.

Table 7.2 Summary output of regressions between sea surface temperature and trace elements for *D. gigas* (Liu 2012)

Elemental combinations	R^2	Standard error	F	p
Sr	0.466	0.781	8.71	0.01451
Sr, Ba	0.841	0.449	23.85	0.00025
Sr, Ba, Mg	0.857	0.451	16.00	0.00097
Sr, Ba, Mg, Na	0.857	0.483	10.50	0.00440
Sr, Ba, Mg, Na, Mn	0.873	0.492	8.23	0.01164
Sr, Ba, Mg, Na, Mn, Cu	0.881	0.521	6.17	0.03224
Sr, Ba, Mg, Na, Mn, Cu, Zn	0.882	0.579	4.28	0.08907
Sr, Mg	0.470	0.820	3.97	0.05758
Sr, Mg, Na	0.489	0.853	2.56	0.12841
Sr, Mg, Na, Mn	0.511	0.893	1.83	0.22768
Sr, Mg, Na, Mn, Cu	0527	0.948	1.34	0.36267
Sr, Mg, Na, Mn, Cu, Zn	0.535	1.030	0.96	0.52999
Sr, Na	0.480	0.812	4.15	0.05271
Sr, Na, Mn	0.480	0.861	2.46	0.13697
Sr, Na, Mn, Cu	0.513	0.891	1.85	0.22476
Sr, Na, Mn, Cu, Zn	0.532	0.944	1.36	0.35497
Sr, Mn	0.466	0.823	3.92	0.05951
Sr, Mn, Cu	0.492	0.851	2.59	0.12574
Sr, Mn, Cu, Zn	0.511	0.893	1,83	0.22837
Sr, Cu	0.480	0.812	4.16	0.05267
Sr, Cu, Zn	0.493	0.850	2.59	0.12491
Sr, Zn	0.466	0.823	3.93	0.05945

There is also a relationship between cephalopod statolith trace elements and salinity. Three sea areas with different salinities, the Arabian Sea, the Indian Ocean, and the Pacific Ocean, differed in the Sr contents of their *S. oualaniensis* statoliths. The statoliths of *D. gigas* distributed in the high salinity Peruvian sea area had significantly higher Sr contents than those distributed in the low salinity Costa Rican ocean area. *Loligo vulgaris*, which lives in relatively high salinity Mediterranean waters, had higher Sr and Zn contents in their statoliths than those in the statoliths of the *Loligo forbesii*, which lives in the lower salinity North Sea. Some studies have found that the statolith of laboratory-reared squid has no significant relationship between Sr and Ba content and salinity, while I/Ca shows a significant negative correlation with salinity.

(3) Migratory Route Reconstruction

The migratory route of *D. gigas* in the Southeastern Pacific Ocean was reconstructed using Sr/Ca and Ba/Ca of statolith as an example.

1) Relationship Between Temperature and Trace Elements

Regression analyses of sea surface temperature (SST) and seven trace elements in the periphery of statolith of *D. gigas* collected at the Chilean offshore showed that Sr/Ca and Ba/Ca combinations were most significantly related to SST with the smallest standard errors (Table 7.2), with the following equation for the relationship: $SST = 33.85 - 0.9996Sr + 0.11040Ba$.

2) Ages and Dates of Ablation Spot in Each Ontogenetic Stage for *D. gigas*

The corresponding age in days of the *D. gigas* migration route selected for spawning in the spring in conjunction with the extrapolation of fishing dates was 18–37 days, 39–73 days, 113–138 days, and 143–176 days, and the corresponding dates extrapolated in conjunction with hatching dates were mainly in October, November, January, and February, respectively (Table 7.3).

Table 7.3 Ages and dates of ablation spot in each ontogenetic stage for *D. gigas* (Liu 2012)

Sample number	Age at each spot corresponding to different ontogenetic stage					Date at each spot corresponding to different ontogenetic stage				
	Embryo	Paralarvae	Juvenile	Subadult	Adult	Embryo	Paralarvae	Juvenile	Subadult	Adult
61a	0	31	58	115	143	2007/9/28	2007/10/29	2007/11/25	2008/1/21	2008/2/18
88a	0	28	61	120	153	2007/9/8	2007/10/6	2007/11/8	2008/1/6	2008/2/8
294a	0	25	65	123	158	2007/11/10	2007/12/5	2008/1/14	2008/3/12	2008/4/16
316a	0	29	73	138	160	2007/9/6	2007/10/5	2007/11/18	2008/1/22	2008/2/13
321a	0	18	39	115	144	2007/10/5	2007/10/23	2007/11/13	2008/1/28	2008/2/26
69b	0	31	55	113	160	2007/9/9	2007/10/10	2007/11/3	2007/12/31	2008/2/16
75b	0	30	53	116	154	2007/10/29	2007/11/28	2007/12/21	2008/2/22	2008/3/31

Table 7.4 Sr/Ca and Ba/Ca ablation spot in each ontogenetic stage for *D. gigas* (Liu 2012)

Sample number	Embryonic phase		Days-old fish		Larval stage (of fish)		Subadult stage (of fishery)		Adult stage	
	Sr/Ca	Ba/Ca	Sr/Ca	Ba/Ca	Sr/Ca	Ba/Ca	Sr/Ca	Ba/Ca	Sr/Ca	Ba/Ca
61a	14.24	10.44	14.53	12.50	14.31	8.79	14.96	19.35	16.21	22.02
88a	14.01	35.37	15.05	12.79	16.78	15.93	18.12	15.92	16.29	19.29
294a	14.85	10.14	14.23	8.34	15.82	13.89	15.23	15.07	16.61	31.26
316a	16.60	12.46	15.88	11.78	15.99	18.01	14.65	16.25	17.57	20.45
321a	16.19	9.35	16.94	14.97	15.55	29.46	16.58	25.35	16.36	15.28
69b	16.79	12.93	15.61	12.91	14.61	15.74	15.58	13.08	16.31	18.95
75b	17.02	12.60	14.58	9.29	16.37	20.06	16.98	20.39	16.79	30.46

3) Trace Elements of *D. gigas* Statoliths at Different Ontogenetic Stage

It was determined that the sample embryonic statolith Sr/Ca ranged from 14.01 to 18.11, and Ba/Ca ranged from 8.75 to 26.57; the paralarval statolith Sr/Ca ranged from 14.01 to 17.13, and Ba/Ca ranged from 9.35 to 35.37; the juvenile statolith Sr/Ca ranged from 14.23 to 16.94, and Ba/Ca ranged from 7.89 to 15.23; the subadult stage statolith Sr/Ca ranged from 12.74 to 16.78, and Ba/Ca range from 8.79 to 29.46; the adult statolith Sr/Ca ranged from 13.38 to 18.12, and Ba/Ca ranged from 12.96 to 25.35 (Table 7.4).

4) Spatial Distribution of *D. gigas* at Different Ontogenetic Stage

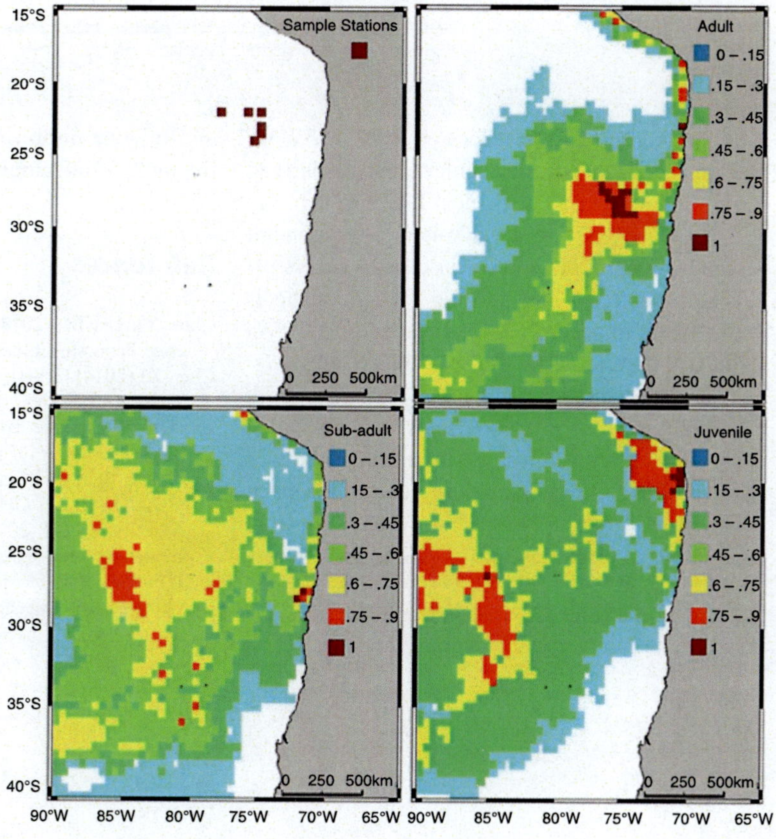

Fig. 7.8 Sample locations and probability distribution of adult, subadult, and juvenile *D. gigas* (Liu 2012)

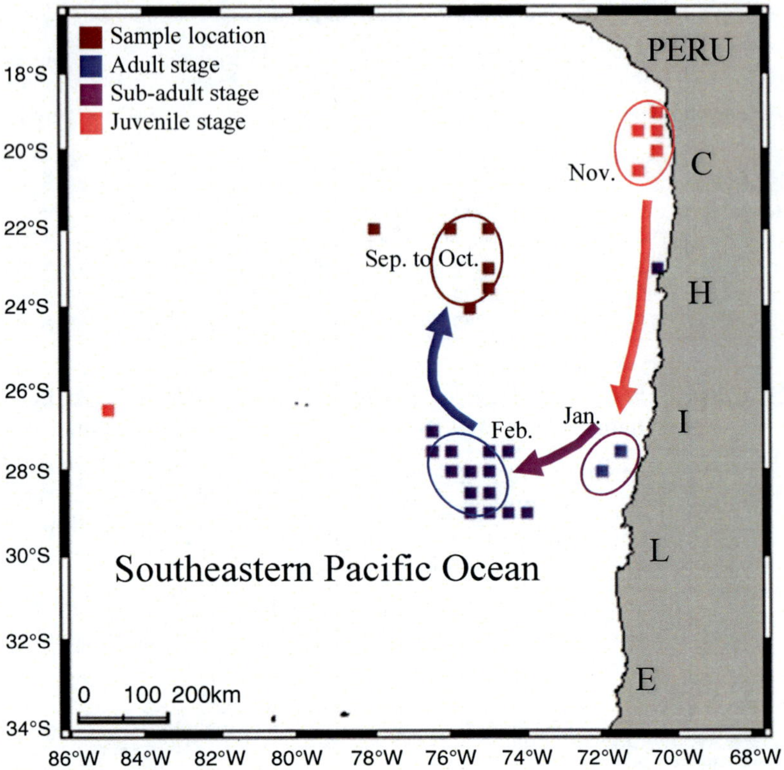

Fig. 7.9 Hypothesized migration routes of *D. gigas* to the Chilean exclusive economic zone based on statolith trace elements (Liu 2012)

The *D. gigas* fishery is located in 74° to 77°W, 22° to 24°S waters; the adult stage, as derived from Sr/Ca and Ba/Ca, is most likely to occur in 74° to 77°W, 27° to 29°S waters; the subadult stage is most likely to occur in coastal waters off 28°S in central Chile; and the juvenile stage is most likely to occur in coastal waters off 20°S in southern Peru to northern Chile (Fig. 7.8).

Linking fishing locations and sea areas with the highest probability of occurrence of adults, subadults, and juveniles, the migration route was projected as follows: juveniles fatten off the coast of northern Chile in November, migrate south to the coast of central Chile at 28°S in January, migrate west to 74°–77°W, 27°–29°S outside the exclusive economic zone (EEZ) in February, and migrate north to 74°–77°W, 22°–24°S from September to October (Fig. 7.9).

References

Chen XJ, Liu BL (2017) Fishery resources biology. Science Press (In Chinese)

Chen XJ (2014) Fisheries resources and fisheries oceanography. Ocean Press (In Chinese)

Liu BL (2012) Life history and population structure of *Dosidicus gigas* off EEZ waters in the Eastern Pacific Ocean. Shanghai Ocean University (In Chinese)

Polovina JJ, Balazs GH, Howell EA, Parker DM, Seki MP, Dutton PH (2004) Forage and migration habitats of logger-head (*Caretta caretta*) and olive ridley (*Lepidochelys olivacea*) sea turtles in the central North Pacific Ocean. Fish Oceanogr 13:36–51